UNITED NATIONS CONFERENCE ON TRADE AND DEVELOPMENT

TECHNOLOGY AND INNOVATION REPORT 2011

Powering Development with Renewable Energy Technologies

UNITED NATIONS
New York and Geneva, 2011

NOTE

The terms country/economy as used in this Report also refer, as appropriate, to territories or areas; the designations employed and the presentation of the material do not imply the expression of any opinion whatsoever on the part of the Secretariat of the United Nations concerning the legal status of any country, territory, city or area or of its authorities, or concerning the delimitation of its frontiers or boundaries. In addition, the designations of country groups are intended solely for statistical or analytical convenience and do not necessarily express a judgment about the stage of development reached by a particular country or area in the development process. The major country groupings used in this Report follow the classification of the United Nations Statistical Office. These are:

Developed countries: the member countries of the OECD (other than Chile, Mexico, the Republic of Korea and Turkey), plus the new European Union member countries which are not OECD members (Bulgaria, Cyprus, Latvia, Lithuania, Malta and Romania), plus Andorra, Bermuda, Liechtenstein, Monaco and San Marino.

Transition economies: South-East Europe and the Commonwealth of Independent States.

Developing economies: in general all economies not specified above. For statistical purposes, the data for China do not include those for Hong Kong Special Administrative Region (Hong Kong SAR), Macao Special Administrative Region (Macao SAR) and Taiwan Province of China.

Least developed countries: These refer to a group of 48 countries that have been identified as "least developed" in terms of their low GDP per capita, their weak human assets and their high degree of economic vulnerability.

The boundaries and names shown and designations used on the maps presented in this publication do not imply official endorsement or acceptance by the United Nations.

Symbols which may have been used in the tables denote the following:

- Two dots (..) indicate that data are not available or are not separately reported. Rows in tables are omitted in those cases where no data are available for any of the elements in the row.
- A dash (–) indicates that the item is equal to zero or its value is negligible.
- A blank in a table indicates that the item is not applicable, unless otherwise indicated.
- A slash (/) between dates representing years (e.g., 1994/95) indicates a financial year.
- Use of a dash (–) between dates representing years (e.g. 1994–1995) signifies the full period involved, including the beginning and end years.
- Reference to "dollars" ($) means United States dollars, unless otherwise indicated.
- Details and percentages in tables do not necessarily add to totals because of rounding.

The material contained in this study may be freely quoted with appropriate acknowledgement.

UNITED NATIONS PUBLICATION
UNCTAD/TIR/2011
Sales No. E.11.II.D.20
ISSN 2076-2917
ISBN 978-92-1-112840-6

PREFACE

As the evidence and impact of climate change increase, so does the urgency to develop new, clean ways of generating and using energy. And as global demand for energy increases, this quest will become even more urgent. This year the population of the planet reached 7 billion. By 2050 it may top 9 billion. All will need access to modern and affordable energy services.

The UNCTAD *Technology and Innovation Report 2011* focuses on the important role of renewable energy technologies in responding to the dual challenge of reducing energy poverty while mitigating climate change. This is particularly timely as the global community prepares for the Rio+20 Conference next year. The Report identifies key capacity issues for developing countries and proposes concrete recommendations for the wider use of renewable energy technologies to promote sustainable development and poverty reduction.

My high-level Advisory Group on Energy and Climate Change stressed that there is an urgent need to mobilize resources and accelerate efforts to ensure universal access to energy. Creating an enabling environment for the promotion and use of renewable energy technologies is a critical part of this effort, as recognized by the United Nations General Assembly when it declared next year as the "International Year for Sustainable Energy for All".

It is also at the heart of my recent launch of the Sustainable Energy for All initiative to help ensure universal access to modern energy services; double the rate of improvement in energy efficiency; and double the share of renewable energy in the global energy mix, all by the year 2030.

We can tackle both energy poverty and climate change by facilitating investment, enhancing access to technologies, and doing more to help developing countries make a transition to a greener path of economic growth. *The Technology and Innovation Report 2011* helps point the way forward.

BAN Ki-moon
Secretary-General
United Nations

ACKNOWLEDGEMENTS

The Technology and Innovation Report 2011 was prepared under the overall direction of Anne Miroux, Director of UNCTAD's Division on Technology and Logistics, and the direct supervision of Mongi Hamdi, Head, Science, Technology and ICT Branch.

The report was written by a team comprising Padmashree Gehl Sampath (team leader), Michael Lim and Carlos Razo. Inputs were provided by Dolf Gielen (Executive Director, IRENA Technology and Innovation Center, Bonn), Professor Mark Jaccard, Simon Fraser University, Professor Robert Ayres (INSEAD), Aaron Cosbey (IISD), Mathew Savage (IISD), Angel Gonzalez-Sanz (UNCTAD), Oliver Johnson (UNCTAD) and Kiyoshi Adachi (UNCTAD).

An ad hoc expert group meeting was organized in Geneva to peer review the report in its draft form. UNCTAD wishes to acknowledge the comments and suggestions provided by the following experts at the meeting: Amit Kumar (The Energy and Resources Institute), Elisa Lanzi (OECD), Pedro Roffe (ICTSD), Ahmed Abdel Latif (ICTSD), Vincent Yu (South Centre), Taffere Tesfachew (UNCTAD), Torbjorn Fredriksson (UNCTAD) and Zeljka Kozul-Wright (UNCTAD). UNCTAD also acknowledges comments by the following experts: Manuel Montes (UN/DESA), Francis Yamba (Centre for Energy Environment and Engineering, Zambia), Aiming Zhou (Asian Development Bank), Youssef Arfaoui (African Development Bank), Mahesh Sugathan (ICTSD), Jean Acquatella (ECLAC) and Alfredo Saad-Filho (UNCTAD).

The report was edited by Praveen Bhalla and research assistance was provided by Fernanda Vilela Ferreira and Hector Dip. Nathalie Loriot was responsible for formatting and Sophie Combette designed the layout.

CONTENTS

Note ii
Preface iii
Acknowledgements iv
Contents v
List of abbreviations x
Key messages xii
Overview xv

CHAPTER I. RENEWABLE ENERGY TECHNOLOGIES, ENERGY POVERTY AND CLIMATE CHANGE: AN INTRODUCTION 1

A. Background 3

B. A new urgency for renewable energies 4
1. An energy perspective 4
2. A climate change perspective 6
3. A developmental perspective 6
4. An equity and inclusiveness perspective 7

C. Energy poverty and a greener catch-up: The role of science, technology and innovation policies 8
1. Towards technological leapfrogging 9
2. The crucial role of technology and innovation policies 10
3. Definitions of key terms
   a. Energy poverty 12
   b. Renewable energy technologies 12

D. Organization of the Report 13

CHAPTER II. RENEWABLE ENERGY TECHNOLOGIES AND THEIR GROWING ROLE IN ENERGY SYSTEMS 19

A. Introduction 21

B. Defining alternative, clean and renewable energies 22
1. The growing role of RETs in energy systems 23
2. Limits of RET applicability 26
3. Established and emerging RETs 28
   a. Hydropower technologies 28
   b. Biomass energy technologies 29
      (i) Traditional biomass 29
      (ii) Modern biomass for electric power 30
      (iii) First and second generation biofuels 30
   c. Wind energy technologies 31
   d. Solar energy technologies 33
      (i) Concentrating solar power systems 33
      (ii) Solar thermal systems 33
      (iii) Solar photovoltaic technology 34
   e. Geothermal energy technology 35
   f. Ocean energy technologies 36
   g. Energy storage technologies 37
4. Scenarios on the future role of RETs in energy systems 38

C. Trends in global investments and costs of RETs **39**
1. Private and public sector investments in RETs 39
2. Costs of renewable energy and other energy sources compared 41
a. Problems with making direct cost comparisons 42
(i) Fiscal support by governments 42
(ii) Factoring in costs specific to conventional energy: Subsidies and environmental externalities 42
(iii) Factoring in costs specific to RETs 42
b. Incorporating costs into the market price of energy options 43
3. The evidence on renewable energy costs 44

D. Summary **48**

CHAPTER III. STIMULATING TECHNICAL CHANGE AND INNOVATION IN AND THROUGH RENEWABLE ENERGY TECHNOLOGIES **53**

A. Introduction **55**

B. Technology and innovation capabilities for RETs development: The context
1. Key networks and interlinkages for RETs 57
a. Public science through public research institutions and centres of excellence 59
b. Private sector enterprises 60
c. End-users (households, communities and commercial enterprises) 62
2. Linkages between RETs and other sectors of the economy 63

C. Promoting a virtuous integration of RETs and STI capacity **64**
1. Addressing systemic failures in RETs 65
2. Tipping the balance in favour of RETs 66
a. Government agencies and the general policy environment 67
b. Facilitation of technology acquisition in the public and private sector 69
c. Promotion of specific renewable energy programmes and policies 69
d. Attainment of grid parity and subsidy issues 69
e. Promoting greater investment and financing options 70
f. Monetizing the costs of energy storage and supply 70
3. Job creation and poverty reduction through RETs 71

D. Summary **72**

CHAPTER IV. INTERNATIONAL POLICY CHALLENGES FOR ACQUISITION, USE AND DEVELOPMENT OF RENEWABLE ENERGY TECHNOLOGIES **79**

A. Introduction **81**

B. International resource mobilization and public financing of RETs **82**
1. Financing within the climate change framework 82
2. Other sources of finance 85
3. International support for financing of RETs: Outstanding issues 87

C. Technology transfer, intellectual property and access to technologies **88**
1. Technology transfer issues within the climate change framework 90
2. Intellectual property rights and RETs 91
a. The barrier versus incentive arguments 92
b. Preliminary trends in patented RETs 95
3. Outstanding issues in the debate on intellectual property and technology transfer 95
a. Beyond technology transfer to technology assimilation 97

b. Assessing the quality – and not the quantity – of technology transfer 97
c. Exploring flexibilities and other options within and outside the TRIPS framework 98

D. The green economy and the Rio+20 framework 98
1. Emerging standards: Carbon footprints and border carbon adjustments 58
2. Preventing misuse of the "green economy" concept 100

E. Framing key issues from a climate change-energy poverty perspective 101
1. Supporting innovation and enabling technological leapfrogging 101
a. An international innovation network for LDCs, with a RETs focus 102
b. Global and regional research funds for RETs deployment and demonstration 103
c. An international technology transfer fund for RETs 103
d. An international training platform for RETs 104
2. Coordinating international support for alleviating energy poverty and mitigating climate change 104
3. Exploring the potential for South-South collaboration 105

F. Summary 106

CHAPTER V. NATIONAL POLICY FRAMEWORKS FOR RENEWABLE ENERGY TECHNOLOGIES 111

A. Introduction 113

B. Enacting policies with RET components and targets 114

C. Specific policy incentives for production and innovation of RETs 117
1. Incentives for innovation of RETs 118
a. Public research grants 119
b. Grants and incentives for innovation of RETs 119
c. Collaborative technology development and public-private partnerships 121
d. Green technology clusters and special economic zones for low-carbon technologies 121
2. Innovation and production incentives and regulatory instruments in energy policies 123
a. Quota obligations/renewable portfolio standards 123
b. Feed-in tariffs 124
3. Flexibilities in the intellectual property rights regime 126
4. Applicability of policy incentives to developing countries 127

D. Adoption and use of new RETs: Policy options and challenges 128
1. Supporting the development of technological absorptive capacity 129
a. Establishing training centres for RETs 129
b. Development of adaptation capabilities 130
c. Education, awareness and outreach 130
2. Elimination of subsidies for conventional energy sources 131
a. Removal of subsidies for carbon-intensive fuels. 131
b. Carbon and energy taxes 132
c. Public procurement of renewable energy 132
3. Applicability of policy incentives to developing countries 133

E. Mobilizing domestic resources and investment in RETs 134
1. Grants and concessional loans 134
2. Tendering systems 135
3. Fiscal measures 135
4. Facilitating foreign direct investment in RETs 135

F. National policies on South-South collaboration and regional integration of RETs................. 137
1. Best practices in trade and investment 138
2. Best practices in technology cooperation 139

G. Summary 140

CHAPTER VI. CONCLUSION 147

List of boxes

Box 1.1: Energy demand and the role of RETs........ 5
Box 1.2: Africa's energy challenge 5
Box 2.1: Definition of renewable energy........ 23
Box 2.2. Developing "smart grids" to better integrate RE sources into energy systems........ 27
Box 2.3: Energy efficiency and conventional measures of thermodynamic efficiency........ 28
Box 2.4: Prices, production and capacity of PV systems 35
Box 2.5. Geothermal energy: technical aspects........ 35
Box 2.6: Ocean energy technologies: technical aspects 36
Box 2.7: Energy scenarios and the future role of RETs........ 38
Box 2.8: Declining costs of RETs: Summary of findings of the IPCC 46
Box 3.1: Policy-relevant insights into technology and innovation........ 57
Box 3.2: The role of public science in building technological and innovative capacity........ 59
Box 3.3: Examples of private firms in wind and solar energy: China and India........ 62
Box 3.4: Constraints on technology and innovation in developing countries 68
Box 4.1: Kyoto Protocol, emissions control and burden sharing 83
Box 4.2: Technology transfer and Article 66.2 of the WTO Agreement on Trade-Related Aspects of Intellectual Property Rights 89
Box 4.3: The Clean Development Mechanism and technology transfer........ 91
Box 4.4: Patents in clean energy: Findings of the UNEP, EPO, ICTSD study 92
Box 4.5: The International Renewable Energy Agency 105
Box 5.1: Demystifying solar energy in poor communities: The Barefoot College in action........ 116
Box 5.2: Lighting a billion Lives: A success story of rural electrification in India 117
Box 5.3: Increasing energy security through wind power in Chile........ 118
Box 5.4: Public support for RETs in the United States of America and China 119
Box 5.5: Examples of grant schemes in industrialized countries 120
Box 5.6: The Renewable Energy and Adaptation to Climate Technologies Programme of the Africa Enterprise Challenge Fund (AECF REACT) 120
Box 5.7: Lighting Africa 121
Box 5.8: Examples of public-private partnerships 122
Box 5.9: Renewable portfolio standards in the Philippines 124
Box 5.10: Feed-in tariffs for biogas and solar PV, Kenya 126
Box 5.11: Promoting integrated approaches for increased production and use of RETs........ 128
Box 5.12: Importance of training for RETs: Experiences of Botswana and Bangladesh 129
Box 5.13: Renewable energy technologies in Asia: Regional research and dissemination programme 130
Box 5.14: Tradable emissions permits 133
Box 5.15: FDI in the global market for solar, wind and biomass 136
Box 5.16: The Turkana Wind Power Project, Kenya........ 139

List of figures

Figure 2.1: Renewable electric power capacity (excluding hydro), end 2010 24
Figure 2.2: Global electricity supply by energy source, 2010 24
Figure 2.3. Levelized costs of some renewable energy technologies compared, 2008 and 2009 ($/MWh) 45
Figure 4.1: Funding arrangements of the UNFCCC 85
Figure 4.2: Sustainable energy financing along the innovation chain 86
Figure 4.3: Number of patent applications for five renewable energy sources, 1979–2003 93
Figure 4.4: Total number of patent applications for energy-generating technologies using renewable and non-fossil sources, 1999–2008 93
Figure 4.5: Market shares held by the first, top five and top ten patent applicants in select technologies
Figure 4.6: Countries' shares of patents in solar thermal and solar photovoltaics, ... 94
1988–2007 (per cent) 94
Figure 4.7: Patent flows in solar PV technologies from Annex I to non-Annex I countries, 1978–2007 96
Figure 4.8: Patent flows in wind power technologies from Annex I to non-Annex I countries, 1988–2007 96
Figure 4.9: Patent flows in biofuel technologies from Annex I to non-Annex I countries, 1988–2007 96

List of tables

Table 2.1: Global investment in renewable energy and related technologies, 2004–2010 ($ billion) 39
Table 2.2: RETs characteristics and energy costs 47
Table 3.1: Relative specializations and potential entry points for firms in wind and solar energies 61
Table 3.2: Access to electricity and urban and rural electrification rates, by region, 2009 64
Table 3.3: Annual investments in global innovation in various energy sources, 2010 ($ billion) 66
Table 4.1: Estimates of RETs investments needed for climate change adaptation and mitigation 83
Table 4.2: Multilateral and bilateral funding for low-carbon technologies 84
Table 4.3: Companies using the United Kingdom's Carbon Reduction Label (by April 2011) 99
Table 5.1: Renewable energy targets of selected developing economies 115
Table 5.2: Countries/states/provinces with RPS policies 123
Table 5.3: Countries/states/provinces with feed-in tariff policies 125

LIST OF ABBREVIATIONS

ACRE	Australian Centre for Renewable Energy
ADB	Asian Development Bank
AECF REACT	The Renewable Energy and Adaptation to Climate Technologies programme of the Africa Enterprise Challenge Fund
AGECC	Advisory Group on Energy and Climate Change (of the United Nations)
ARPA-E	Advanced Research Project Agency-Energy
BCA	border carbon adjustment
BSE	barefoot solar engineers
Btu	British thermal unit
CCGT	combined cycle gas turbine
CCL	climate change levy
CCMT	climate change mitigation technologies
CCS	carbon capture and storage
CDIP	Committee on Development and Intellectual Property
CDM	Clean Development Mechanism
CET	clean energy technology
CHP	combined heat and power
CIS	Commonwealth of Independent States
CO_2	carbon dioxide
COP	Conference of the Parties
CSIR	Council for Scientific and Industrial Research
CSP	concentrating solar power
CTF	Clean Technology Fund
EC	European Commission
EGS	engineered thermal system
EJ	exajoule
EPO	European Patent Office
ESMAP	Energy Sector Management Assistance Programme
EST	environmentally sound technology
ETI	Energy Technologies Institute
EU	European Union
EU ETS	European Union's Emission Trading System
EU-27	27 member countries of the European Union
FDI	foreign direct investment
FIT	feed-in tariff
GDP	gross domestic product
GEF	Global Environment Facility
GHG	greenhouse gas
GW	gigawatt
GWEC	Global Wind Energy Council
G-20	group of 20
ICT	information and communication technologies
IEA	International Energy Agency
IFC	International Finance Corporation
IGCC	integrated gasification combined cycle
IITC	IRENA Innovation and Technology Centre
IPCC	Intergovernmental Panel on Climate Change
IPR	intellectual property right
IRENA	International Renewable Energy Agency

JPOI	Johannesburg Plan of Implementation
kgoe	kilogram of oil equivalent
km	kilometer
KMTC	Knowledge Management and Technology Cooperation
kW	kilowatt
kWh	kilowatt-hour
kWp	kilowatt-peak
LaBL	Lighting a billion Lives
LCOE	levelized cost of electricity
LDC	least developed country
LDCF	Least Developed Country Fund
MDG	Millennium Development Goal
MW	megawatt
MWh	megawatt-hour
MWp	megawatt power
NGO	non-governmental organization
NRI	national research institution
OECD	Organization for Economic Co-operation and Development
OTEC	ocean thermal energy conversion
PCT	Patent Cooperation Treaty
PPA	power purchase agreement
PPM	particles per million
PPP	public private partnership
PV	photovoltaic
R&D	research and development
RE	renewable energy
REDP	Renewable Energy Demonstration Program
RET	renewable energy technology
ROC	renewables obligation certificate
RPS	renewable portfolio standard
SCCF	Strategic Climate Change Fund
SEZ	special economic zone
SHS	solar home systems
SIDA	Swedish International Development Agency
SME	small and medium-sized enterprise
STI	science, technology and innovation
TERI	The Energy and Resources Institute
TIR	*Technology and Innovation Report*
TRIPS	Trade-related Aspects of Intellectual Property Rights (WTO Agreement)
TWh	terawatt-hour (a terawatt is equivalent to a million megawatts)
UNCSD	United Nations Conference on Sustainable Development
UNCTAD	United Nations Conference on Trade and Development
UN/DESA	United Nations Department of Economic and Social Affairs
UNDP	United Nations Development Programme
UNEP	United Nations Environment Programme
UNFCCC	United Nations Framework Convention on Climate Change
UNIDO	United Nations Industrial Development Organization
USAID	United States Agency for International Development
VAT	value-added tax
W	watt
WIPO	World Intellectual Property Organization
WSSD	World Summit on Sustainable Development
WTO	World Trade Organization

KEY MESSAGES

On technology and innovation capacity for RETs:

1. A mutually compatible response to the dual challenge of reducing energy poverty and mitigating climate change requires a new energy paradigm. Such a paradigm would have RETs complementing (and eventually substituting) conventional energy sources in promoting universal access to energy.
2. Established RETs, such as solar PV technologies and onshore wind, are experiencing rapid ongoing technological progress and reductions in energy generation costs.
3. RETs are already being deployed on a significant scale in some countries, though this varies by region.
4. Much progress can be achieved in alleviating energy poverty by focusing on rural, off grid applications alongside efforts to establish more technologically and financially intensive grid-based RET applications.
5. In the absence of technological capabilities, national strategies for sustainable economic development are likely to be constantly undermined.
6. Strengthening technological absorptive capacities is essential not only to build R&D capabilities for RETs in the short and mid term, but also to promote adaptation and dissemination of RETs.
7. RETs use should be integrated within broader goals for poverty reduction and job creation for the more economically vulnerable groups in developing country economies.

On the international policy challenges for RETs:

1. There is an urgent need to reposition the debate within the international agenda on climate change so that obligations of countries to mitigate climate change is framed in terms of creating development opportunities for all in an environmentally sustainable manner.
2. Such a repositioning also implies focusing on issues of finance, technology transfer and technology dissemination for developing countries in the context of RETs.
3. The current international finance and technology transfer architecture is fragmented. It needs to be strengthened with the aim of reducing energy poverty while mitigating climate change.
4. International support needs to work hand in hand with national frameworks on RETs, complementing efforts in three critical areas: increasing financial resources for RETs, promoting greater access to technology and enabling greater technological learning within the green economy and the Rio-plus-twenty framework.
5. The diffusion of RETs in developing countries involves much more than transferring technology hardware from one location to another. This Report, noting the complexity of technological change in different contexts, calls for targeted international support to foster RETs-related learning. Such support could include the following elements:
   (i) an international innovation network for LDCs, with a RET focus, that seeks to facilitate knowledge accumulation and innovation in LDCs.
   (ii) global and regional research funds for RETs deployment and demonstration, that focus attention on making resources available to adaptation and incremental innovations in RETs for use in a wide variety of contexts.
   (iii) an international RETs technology transfer fund that is dedicated to facilitating private-private and private-public transfer of technology for RETs.
   (iv) an international RETs training platform that promotes capacity building and skills accumulation in developing countries.
6. More support could take the form of augmenting and further strengthening the recently proposed technology mechanism within the UNFCCC, particularly by increasing its focus on RETs.

On national policy frameworks for RETs:

1. National governments in developing countries can play a pivotal role in combining conventional sources of energy with RETs in ways that will not only help reduce energy poverty, but also simultaneously promote climate-friendly solutions to development.
2. This Report proposes that developing countries adopt a national integrated innovation policy framework to create policy incentives in national innovation policies and national energy policies for the greater use, diffusion, production and innovation of RETs.
3. Such a policy framework would have five key functions:
   (i) Defining policy strategies and goals;
   (ii) Providing policy incentives for R&D, innovation and production of RETs;
   (iii) Providing policy incentives for developing greater technological absorptive capacity, which is needed for adaptation and use of available RETs;
   (iv) Promoting domestic resource mobilization for RETs in national contexts; and,
   (v) Exploring newer means of improving innovation capacity in RETs, including South-South collaboration.
4. Not all of the policy options proposed in the Report are available or applicable to all developing countries and LDCs.
5. Incentives for RETs production and innovation can be entrenched into the wider innovation policy framework and energy policies of countries through a variety of policy measures.
6. For the poorer countries, the ability to undertake large-scale R&D or establish significant manufacturing capacity will be constrained by the relatively small size of their domestic markets, lack of access to finance and weak institutional capacity. In such cases, countries should consider incentives to build greater absorption capacity in RETs and revisit their energy subsidy policies.
7. Incentive structures can start small, on low-scale projects, designed to encourage private sector solutions to renewable energy technology development and deployment challenges in rural settings.
8. Developing countries will face different problems in RETs promotion, production and innovation, depending on their respective starting points. Nevertheless, for all developing countries, RETs present real opportunities for reducing energy poverty, and the right policies could influence the extent of benefits that could be derived from RETs use, adaptation and dissemination.

OVERVIEW

RENEWABLE ENERGY TECHNOLOGIES, ENERGY POVERTY AND CLIMATE CHANGE

Sustained economic growth of the kind that leads to continuous improvement in the living standards of all people through poverty reduction rests on assuring access to energy for all. Such a global energy access agenda requires a greater focus on energy generation and use from existing resources while minimizing waste. However, the use of conventional energy sources (primarily fossil fuels) are believed to have led to a rise in greenhouse gas (GHG) emissions and to a resulting increase in global average temperatures since the mid-twentieth century. The fundamental conclusions of the most recent assessment report of the IPCC are that climate change is the result of human activity, that the ongoing rate of climate change will have devastating effects if left unchecked, and that the costs of action for mitigation and adaptation would be significantly lower than the costs of inaction.[1] Therefore from a climate change perspective, there is a need for all countries worldwide to embark upon low-carbon, high growth trajectories. It also requires promoting the use of other, newer or more cost-effective energy sources in all countries, which could complement the conventional energy supplies predominantly in use today. Renewable energy (RE) sources offer one such distinct possibility, and established renewable energy technologies (RETs) can complement more traditional sources of energy, thereby providing countries with varied energy options within their national energy matrices to suit their specific needs and conditions. Given their enormous potential, there is growing interest in the current and future role of RETs in national energy supply systems worldwide.

This, however, is not an easy task for all developing and least developed countries (LDCs), since the greater use of RETs for energy supply and industrial development is dependent on building technological capabilities. Against this broad background, the *Technology and Innovation Report (TIR) 2011* analyses the important role of technology and innovation policies in expanding the application and wider acceptance of RETs, particularly in the context of developing countries. Technology and innovation policies can promote and facilitate the development, acquisition, adaptation and deployment of RETs to support sustainable development and poverty reduction in developing countries and LDCs.

Four current trends lend a new urgency to the need to explore how far and how easily RETs could serve energy needs worldwide. First, ensuring universal access to conventional energy sources using grids entails high costs, which means that developing countries are unlikely to be able to afford the costs of linking additional households, especially those in rural areas, to existing grids. Second, the climate change debate has injected a greater sense of urgency into searching for newer energy options, as a result of both ongoing policy negotiations and the greater incidence of environmental catastrophes worldwide. Third, from a development perspective, the recent financial and environmental crises have caused major setbacks in a large number of developing countries and LDCs, resulting in their further marginalization from the global economy. The LDCs and many developing countries suffer from severe structural vulnerabilities that are a result of their patterns of integration into the global economy. The international community needs to promote low-carbon, climate-friendly development while fostering inclusive economic growth in these economies as a matter of urgency. Lastly, there are extreme inequalities within developing countries themselves, and lack of access to energy affects the poorest of the poor worldwide, impeding their ability to enjoy the basic amenities of modern life that are available to others at the same level of development.

Within the United Nations Framework Convention on Climate Change (UNFCCC), polarized positions on who should shoulder responsibility for the current state of emissions and share the financial burden for mitigating climate change are based on the erroneous belief in the incompatibility of the dual challenges of promoting industrial development and mitigating climate change. Developing countries, in particular, face the challenge of promoting industrial development – a fundamental prerequisite for poverty reduction and equality in their societies – while reducing their reliance on conventional energy sources that have played a central role in global economic growth until recently.[2] Most of these countries also remain far more vulnerable to most of the environmental threats arising from climate change.[3]

However, the advantages of using RETs will not accrue automatically in developing countries. Although many of the RETs needed for meeting a larger share of the global energy demand already exist, or are on the verge of commercialization, the knowledge and technological capabilities required for their transfer to developing countries and LDCs are not easily accessible. Developing countries will need to strengthen their innovation systems,[4] policy frameworks and linkages to enable wider RET dissemination and to promote a greener catch-up process. Promoting greater access to RETs and support for use and adaptation of these technologies through all means possible will be important to enable developing countries to sustainably integrate these processes into efforts aimed at capital formation and transformation of their productive structures.

There is a need not only for strong domestic technology and innovation policies, but also for greater international efforts to make the international trade and intellectual property rights (IPRs) regime more supportive of the technological needs of developing countries and LDCs. International support to developing countries through various channels should also include financial support and North–South, South–South and triangular cooperation, as well as effective technology transfer mechanisms. All of these will be necessary complements to the development of local capacities for RETs.

This *TIR* identifies five distinct issues that stand out in the debate on technology and innovation of RETs, which are of particular relevance for all developing countries and LDCs. First, structural transformation that supports the economic development of countries relies strongly on the growth of national technological capabilities. At present, inadequate energy supply is a constraint that applies not only to the manufacturing sector, but also to other sectors that are potentially important to the process of industrialization and development, such as services, tourism and agricultural processing, which depend on reliable, high-quality power supply. It is therefore important to recognize the virtuous relationship between energy security and technological capabilities: energy security is a key aspect of the physical infrastructure required for growth, and technological capabilities are a fundamental prerequisite for greater adaptation and use of RETs within domestic economies.

Second, incoherent, and often conflicting, policy developments at the multilateral level tend to adversely affect national aspirations for technological empowerment in developing countries. Although climate change will affect all countries and communities worldwide, developing countries (especially LDCs) will shoulder a disproportionate burden from the fallout resulting from climate change, including increasing climatic variations, extreme weather events and natural disasters. The ongoing debates on climate change reflect the diverse positions of countries on how the burden should be shared.

Third, the issue of greater transfer of climate-friendly technologies that has been a key element in the global debate on climate change is intricately linked to technology and innovation infrastructures in countries. In the renewable energy (RE) sector, recent evidence shows that basic approaches to solving technological problems have long been off-patent, and therefore can be adapted and disseminated in developing countries *provided* that some technological prerequisites are met. This points to the need for greater attention to strengthening the technology absorptive capacity of countries through coordinated policy support, in addition to making existing technologies available and assisting in their greater diffusion.

Fourth, RETs will remain a distant goal as long as they are prohibitively expensive. Innovation in RETs is moving at a fast pace globally, but left on its own, or left to the "market", it is unclear to what extent this pace will continue globally and to what extent it will lower the prices of these technologies for use at the individual household and firm level in the medium term.

Finally, RETs form part of the wider debate on emerging patterns of investment and technology that fall under the umbrella of the "green economy". At a fundamental level, the concept of the "green economy" itself has been highly contested. Some argue that calling for large-scale investments in developing countries to facilitate the transition to a green economy imposes uneven costs, thereby creating an additional burden on already disadvantaged groups of people. The challenge is to ensure that the green economy concept, which will also be the focus of the Rio+20 framework, is structured in a way that it does not adversely affect ongoing productive activities in developing countries while helping their transition to "green" modes of development. Numerous issues will need to be addressed in this context, including patterns of trade, technological upgrading and specialization.

Analysing these five issues at length, the *TIR 2011* argues that RETs can bring numerous benefits to developing countries. The potential impacts of RETs in terms of reducing energy poverty, generating employment and creating new production and innovative activity add to their environmental advantages. Several established RETs have significant potential to contribute to a broad range of development goals. It is beyond the scope of this Report to address the whole range of policy implications of all RETs in the very different contexts of the various categories of developing countries. It therefore focuses on those that are (i) already mature enough to make practical contributions to policy objectives in the short term, but are sufficiently recent in their commercialization to present challenges with which policymakers may be less familiar; and (ii) particularly appropriate to the objective of reducing and eventually eliminating energy poverty in developing countries as complements (and eventually substitutes) to conventional energy sources.

THE EXPANDING ROLE OF RENEWABLE ENERGY TECHNOLOGIES

Several RETs are well established

RETs are a diverse group of technologies that are currently at different levels of maturity. Those based on wind, geothermal, solar thermal and hydro are mature technologies and are already being deployed widely. Others, including second-generation biofuels and ocean energy, remain at varying stages of pre-commercial development. Although there are problems of intermittency associated with some of them (for example, in the provision of solar energy, where the sun is available for only a limited number of hours per day), they are very versatile in that they can be deployed in various configurations, either alone or, often, in combination with conventional energy technologies. Therefore they offer the potential to contribute significantly to alleviating energy poverty in diverse situations.

The *TIR* focuses primarily on RETs based on wind, solar and modern biomass sources for electricity generation, either in centralized or decentralized facilities. These are among the most important and fastest growing RETs in developing countries. There are also non-electric applications of REs, such as biofuels that are used for transportation, space heating, hot water and cooking (e.g. by solar cookers).

The role of RETs in alleviating energy poverty is growing

On a global scale, although the various advantages of RETs are increasingly being recognized, established fossil fuel sources still dominate energy supply at present, providing up to 89 per cent of all global energy. In 2008, RE sources (including large hydro installations) accounted for 12.9 per cent of global primary energy supply, whereas the bulk was supplied by fossil fuels (including oil, gas and coal). However, a large proportion of the global population cannot afford these conventional energy supplies. According to estimates of the International Energy Agency (IEA), over 20 per cent of the global population (1.4 billion people approximately) lacked access to electricity in 2010. South Asia has the largest absolute numbers of people without such access (42 per cent of the world total), in spite of recent rapid progress. Taking the entire population of this subregion, 38 per cent have no access to electricity, and within this figure, 49 per cent of people living in rural areas lack access. In relative terms, sub-Saharan Africa is the most underserved region, with 69.5 per cent of the population having no access to electricity, and only a meagre 14 per cent of the rural population having access.

Eliminating energy poverty and promoting greater access to energy to promote economic development therefore requires serious consideration of how RETs could complement and/or even substitute conventional energy sources. Will such a new energy paradigm that envisages a greater role for RETs be able to create greater employment? Could those RETs be deployed in remote rural areas that are hard to connect to the conventional energy grid? Will such RETs be applicable and easy to use by individual users, but at the same time have the potential for scale-up within enterprises, firms and sectors? Would they alleviate, at least partially, the difficulties faced by vulnerable social groups affected by poverty (e.g. rural populations, women, children and indigenous groups) so that they can devote more time and attention to income-generating and other activities?

A significant aspect of RE use is that they offer the possibility of devising semi-grid or off-grid rural installations that promote greater access to energy in developing countries than that provided by conventional energy sources which rely extensively on grid connections. Of the 1.4 billion people not connected to electricity grids globally, approximately 85 per cent live in rural areas. Because of their possibility of use in non-grid or semi-grid applications, RETs can be an important means of energy supply in areas where other energy sources are not available, such as in isolated rural communities. Such decentralized, off-grid applications of RETs are already in relatively wide use in developing countries, where they provide cost-effective energy solutions that bring significant benefits to local communities.

RETs such as solar pumps, solar PV installations, small wind, mini-hydro and biomass mini-grids offer higher potential and cost advantages than traditional grid extension. They can be a reasonable option for providing some degree of access to energy, particularly in rural areas in developing countries and LDCs where national energy grids are unlikely to expand in the near future. Arguably, some of these applications are small in scale and do not make much of an impact on energy provision at the national/global level, but they can still play an important role in reducing energy poverty at the local/rural level. In these cases, RETs offer a realistic option for eradication, or at least for alleviation, of energy poverty.

Technological progress and greater investments and deployment are lowering costs of established RETs

There has been rapid ongoing technological progress in some RETs, such as solar PV technologies and onshore wind energy, with accompanying reductions in energy generation costs. The cost competitiveness of RETs relative to conventional energy sources is also improving, and can be expected to improve even further with continued technological progress and higher investment in their development, production and deployment. The prices of solar PV systems, for instance, have been falling extremely rapidly. During the 18 months leading to June 2010, prices of new solar panel modules fell by an estimated 50 per cent. And in some off-grid and mini-grid applications some RETs were already competitive with conventional energy in 2005, even with the relatively low oil prices prevailing at that time. It is reported that in Africa, Asia and Latin America, the demand for modern energy is driving the use of PV for mini-grid or off-grid solar systems, which in many instances are already at price parity with fossil fuels. This implies that for precisely those applications which may be most suitable for isolated communities (i.e. decentralized applications that do not require connection to the national or regional energy grids) RETs may be at their most cost-competitive. Rising, and increasingly volatile, oil prices and growing investments in RETs may also be contributing to this trend. However, additional technological improvements that could help to better integrate RE into the existing energy infrastructure (including through the development of smart energy grids) and augment the storage capacities of RETs will be valuable in promoting their cost competitiveness.

Despite the ongoing surge in the deployment of RETs, at present they account for only a small fraction of global energy consumption. The *TIR 2011* stresses that there is still enormous technical potential for power generation from RETs, and argues that such RETs are likely to play an increasingly important role in meeting global energy demand as continued technological progress, additional investment and further deployment lead to cost reductions over the medium and long term globally. The analysis in the Report shows that RETs will continue to evolve as complements to existing energy sources globally, with the eventual aim of replacing conventional energy in the long term. For developing countries and LDCs, this is a positive trend. The actual speed and extent of deployment of RETs and the role they will eventually play will depend critically on the policy choices that are made today and in the future. The policy issues that need to be considered within national frameworks for technology and innovation and the ways and means of international support will be critical for harnessing the potential of RETs for poverty reduction and sustainable development.

STIMULATING TECHNICAL CHANGE AND INNOVATION IN AND THROUGH RENEWABLE ENERGY TECHNOLOGIES

Technology and innovation capacity and reliable energy supply are intricately linked

Uninterrupted and reliable energy supply is an important stimulant to innovative capacity and economic growth. Indeed, a number of studies underline a direct causal relationship between the low supply of electricity and stunted economic growth. At the same time, technology and innovation capabilities are important for promoting R&D and innovation to produce state-of-the-art RETs, and for creating a critical base of knowledge that is essential for adapting and disseminating RETs. A critical threshold of technological capability is also a prerequisite for making technical improvements to RETs that enable significant cost reductions so that they can be deployed on a larger scale in developing countries. The success of RETs-related technology transfer initiatives also depends on the ability of actors in developing countries to absorb and apply the technologies transferred. The absence of, or limitations in, technological and innovation capabilities is therefore likely to constantly undermine national strategies for sustainable development based on the greater use of RETs. This virtuous relationship between RETs and technology and innovation capacity needs to be recognized and fostered actively.

Countries' capacities for technological absorption need to be strengthened through coordinated policy support, but an additional priority will be to make existing technologies available and assist in their greater diffusion. As noted earlier, while innovation in RETs is moving at a fast pace globally, ensuring this continues will require policies that promote the wider adaptation and deployment of RETs. In the context of the current state of underdeveloped energy infrastructure in developing countries and LDCs, RETs could not only help to reduce energy poverty in many novel ways; they could also help reduce social inequalities through the creation of new jobs associated with the application of RETs. Public policy therefore has an important role to play in this regard, in addition to tipping the balance towards energy mixes that give prominence to RETs development in developing countries.

Innovation policy frameworks for RETs are a fundamental requirement

Innovation systems in developing countries are fundamental to shaping the capacity for the technological learning needed for adaptation, use, production and R&D-based innovation of RETs. There are several features of technology and innovation unique to RETs compared with other sectors that have been the focus of many policy studies. First, there is already a well-established energy system globally, and RETs are technologies that seek to provide alternative solutions to achieve the same results using natural and renewable resources of different kinds (such as sun, wind and water). Their unique selling point is that they offer environmentally friendly solutions to energy needs for the same service, namely the supply of energy. This is different from innovation in other sectors where competition is structured around the provision of newer products and services at reasonable prices.

Second, the intermittency issues related to RETs necessitate a systemic approach to promoting innovation in the sector. Evidence shows that intermittency of different RE supplies can be dealt with quite easily within electricity systems when solutions are designed from a systemic perspective.[5] A systemic treatment of RETs is also important from another perspective, namely, the management of demand for energy. The end-use dimension (i.e. how many people can access a particular supply and how effectively it can be provided) will need to play a major role when considering RETs as a means of alleviating energy poverty in developing countries. Thus a systemic perspective should give due consideration to the demand dimension when designing on-grid, off-grid or semi-grid applications using RETs.

Third, it is often assumed, incorrectly, that technological capability is required primarily for R&D aimed at the creation or development of newer RETs. As the *TIR 2011* shows, technology and innovative capability is also fundamental for other aspects, such as:

(i) Making minor technical improvements that could enable significant cost reductions in production techniques, adaptation and use; and
(ii) Adaptation, dissemination, maintenance and use of existing RETs within key sectors of the economy, which depend not only on the availability of materials, but also on diverse forms of knowledge.

Fourth, in developing countries, there is an urgent need to promote choices in innovation and industrial development based on RETs. These choices may be different depending on the conditions in the country and the kind of RE resource(s) available. The specific characteristics of different RETs, varied project sizes and the possibilities for off-grid and decentralized supply, imply many new players, both in project development (new and existing firms, households and communities) and in financing (existing lenders, new microcredit scheme, government initiatives).

Therefore, strengthening national frameworks for technology and innovation in developing countries is a necessary pre-condition for ensuring increased use and innovation of RETs through: (a) the greater integration of RETs within socio-economic development strategies of countries; (b) creation of capacity for increased technology absorption in general, and in RETs in particular; and (c) express policy support aimed at significantly integrating RETs into the national energy mix by tipping the balance in favour of RETs development, production and use.

National governments need to tip the balance in favour of RETs

There is an urgent need for government action aimed at substituting patterns of current energy use with reliable, established RETs. While off-grid RETs (especially modern biomass-based ones) may be relatively easy to deploy, many still remain very expensive at the scales required to make an impact in developing countries, despite rapid technological advances. For example, a study by the IEA (2009) came to the conclusion that in the United States, electricity from new nuclear power plants was 15–30 per cent more expensive than from coal-fired plants, and the cost of offshore wind power was more than double that of coal, while solar power cost five times as much. Changing from the current global situation of no energy, or unreliable and often undesirable sources of alternative energy (such as traditional biomass), to one where industrial development begins to pursue a cleaner growth trajectory is essential for driving down the costs of RETs.

Each time investment is made in generating more energy through RETs, there is not only a gradual shift in the energy base; it also has a significant impact on the capacity of RETs to supply energy more economically. For example, according to recent reports, every time the amount of wind generation capacity doubles, the price of electricity produced by wind turbines falls by 9–17 per cent.[6] This holds true for all RETs: with each new installation, there is learning attached as to how the technology can be made available more effectively and efficiently in different contexts so as to lower costs over a period of time.

Government action will need to focus on two very important areas of intervention: addressing systemic failures in RETs, and tipping the balance away from a focus on conventional energy sources and towards RETs. Systemic failures in the RETs sector are varied and emerge from sources other than just the market. They can be caused by technological uncertainty, environmental failures or other systemic factors. Therefore, it will be important for government intervention to address these failures.

Policy incentives, critical for inducing a shift towards the wider application of RETs in the energy mix of countries, need to be designed and articulated at the national and regional levels so that collective actions can be fostered. Most importantly, energy production should cater to local needs and demand in countries, for which a systemic perspective is necessary. Policy support needs to be directed at mobilizing greater domestic resources to foster RETs development and use, in addition to providing increased access to the most advanced, cost-cutting technological improvements to established RETs.

Governments can play a vital role in making RETs feasible at each level: use, adaptation, production and innovation. Government agencies and the policy framework should aim at:

(i) Promoting the general innovation environment for the development of science, technology and innovation;
(ii) Making RETs viable; and
(iii) Enabling enterprise development of and through RETs.

This requires governments to adopt an agenda of proactively promoting access to energy services of the kind that is conducive to development, while also focusing on the important positive relationship between technology and innovation capacity and increased use of RETs. Greater international support for developing countries will be critical on both these fronts.

INTERNATIONAL POLICY CHALLENGES FOR ACQUISITION, USE AND DEVELOPMENT OF RENEWABLE ENERGY TECHNOLOGIES

The international discourse needs to be framed more positively, with a focus on mitigating climate change and alleviating energy poverty

Efforts at the national level aimed at harnessing the virtuous relationship between RETs-related technology and innovation capacity for inclusive economic development and climate change mitigation need to be strengthened through greater international support. At the international level, discussions and negotiations on climate change and the green economy have gained momentum in recent years. A major focus of those discussions relates to environmentally sustainable technologies, or low-carbon, "clean" technologies,[7] as a means of contributing to climate change mitigation and adaptation globally.[8] This is a very important global goal, which will serve the needs of developing countries in particular, given the evidence that climate change is having disproportionately damaging impacts on those countries. However, along with efforts to mitigate climate change, there needs to be an equally important focus on eliminating energy poverty in developing countries, not only to improve people's living conditions but also to boost economic development.

The *TIR 2011* stresses upon the need for repositioning issues within the international agenda, whereby the obligations of countries to mitigate climate change are framed in terms of creating development opportunities for all in an environmentally sustainable manner. Central to this repositioning is the triangular relationship between equity, development and environment. From this perspective, recognition of the right of all people worldwide to access energy services is long overdue and needs to be addressed. Developing countries, especially the least developed, have experienced a particularly large share of natural disasters, such as hurricanes, tornados, droughts and flooding, as a result of changing climatic conditions. According to recent estimates, 98 per cent of those seriously affected by natural disasters between 2000 and 2004 and 99 per cent of all casualties of natural disasters in 2008 lived in developing countries, particularly in Africa and South Asia where the world's poorest people live.

Such a repositioning also implies a greater focus on three key challenges, namely international resource mobilization for RETs financing; greater access to technology through technology transfer and the creation of flexibilities in the IPRs regime; and promoting wider use of RETs and technological learning in the push for a green economy and within the Rio+20 framework. These issues have been and remain central to all debates and decisions of the UNFCCC and the Kyoto Protocol that focus mainly on environmentally sustainable – or clean – technologies, of which RETs form a subset. In highlighting the need for a greater focus on RETs in international discussions, the Report also identifies the main hurdles in all these three policy areas.

International financial support for RETs needs to be strengthened and targeted

A number of estimates have been produced that try to quantify the challenge of adaptation and climate change mitigation. All of them consider slightly different categories of investments that will be needed in the immediate or medium term. The International Energy Agency estimate covers only electricity generation technologies, and therefore excludes investment in transport fuels and heating technologies. While all the estimates are indicative, the definitions of technology and the broad goals assumed in the IEA (2000) are probably the most relevant to the issues under consideration in this *TIR*.[9] The proposal to halve energy-related emissions by 2050 corresponds roughly to the minimum mitigation levels deemed necessary by the IPCC, and the definition of low-carbon energy technologies covers RETs. The IEA's estimates for the level of investments needed are lower than the other estimates in the medium term, at \$300–\$400 billion per annum up to 2020, but rise thereafter to reach \$750 billion by 2030.

This raises questions about the capacity of public finance to support the rapid and widespread deployment of RETs as part of adaptation efforts and the role of international support. There are a number of known sources of finance at the multilateral and regional levels such as the World Bank's Climate Investment Funds, the Clean Technology Fund and the newly announced UNFCCC Green Climate Fund. However, several caveats apply when calculating the amount of finances available under all the funding figures. Some of the funds are multiyear commitments and often cover mitigation and adaptation. Also, some of the funds are not yet available. Taking all these caveats together, the total amount of annual funding for RETs from public sources is likely to be about \$5 billion from the known sources. This figure is far from sufficient when compared to the global needs. The International Renewable Energy Agency estimates that just the African continent would need an investment of \$40.6 billion per year to make energy access a reality in a sustainable way.

It is therefore that in the area of finance, support for greater investment in RETs and their use in developing countries is critical today. In response to the global financial and economic crisis, many countries initiated stimulus packages that included funding for efforts to build capacity in those areas of the green economy that display the greatest growth potential. No doubt, the general trend is towards policies that simultaneously aim at securing environmental benefits through increased use of RETs, development benefits through increased energy provision, and economic benefits by increasing domestic capacity in areas that show growth potential.

However, such ongoing efforts in developing countries would be better served if outstanding issues relating to international financial support for RETs could be urgently resolved with the aim of promoting greater innovation, production and use of such technologies. At present, international financing of clean technologies, which is largely multilateral, is highly fragmented, uncoordinated and lacks transparency. It is also woefully inadequate to meet total funding requirements for climate change mitigation and adaptation. While such financing may partly be targeted at RETs, additional international funding for RETs is required as a priority. Coordination of funding sources with the aim of mainstreaming RETs into national energy systems globally should be an important aspect of climate change mitigation efforts. This would not only lead to the development of more efficient energy systems globally; it would also ensure that the financing contributes to greater technological progress towards newer and/or more cost-effective RETs.

Access to RETs and related technology transfer need to be more clearly articulated

Currently, most of the clean technologies needed for developing countries and LDCs are off-patent. Despite this general finding, recent trends show that patenting activity in RETs is on the rise. Following an analysis of these trends against the backdrop of the ongoing negotiations on the draft UNFCCC,[10] the *TIR 2011* suggests that discussions on technology transfer of RETs within the climate change framework should move beyond a narrow focus on the issue of technology transfer to a broader focus on enabling technology assimilation of RETs. Indeed, the recent Climate Change Conference at Cancun in 2010 proposed to strengthen the focus on technology transfer, including the creation of a new technology mechanism to help enhance the technological capacity of countries to absorb and utilize RETs.

Accumulation of technological know-how and learning capabilities is not an automatic process. Learning accompanies the acquisition of production and industrial equipment, including learning how to use and adapt it to local conditions. In order to foster broader technology assimilation, the technology transfer exercise will need to take into account the specific technological dimensions of RETs as well as the nature of actors and organizations in developing countries. The quality of technology transfer should be assessed by the extent to which the recipient's know-how of a product, process or routine activity is enhanced, and not just by the number of technology transfer projects undertaken. A greater articulation of flexibilities under the global IPRs regime in the specific context of RETs is also required.

The green economy and the Rio+20 framework should promote wider use and learning of RETs

In addition to providing critical infrastructure to support the emergence and shift in production structures in developing countries, RETs can serve the goals of their industrial policy by helping those countries' exporters become more competitive in the face of increasingly stringent international environmental standards. However, simply forcing developing countries to use RETs through measures such as carbon labelling and border carbon adjustments may not be sufficient to enable the transition. Indeed, such measures may even have adverse effects on industries in developing countries by acting as barriers to imports, since enterprises and organizations may not have the means (financial and technological) to conform to the new requirements. To ensure that "green" requirements do not place an additional burden on industries in developing countries and LDCs, global efforts aimed at climate change mitigation need to be accompanied by international support in finance and technology to help these countries transition to RETs in a strategic and sustainable manner.

Targeted international mechanisms for RETs-related innovation and technological leapfrogging are required

The obvious question for all developing countries and for the global community is whether the BRICS countries (Brazil, the Russian Federation, India, China and South Africa) are special cases. To some extent they are: they have the prerequisites for competitive production of many RETs, such as a workforce with advanced technical training, supporting industries and services in high-tech areas, access to finance, ample government assistance and a large domestic market, all of which seem to favour these larger emerging developing countries over smaller, poorer developing countries and LDCs.

Historically, promoting technological learning and innovation has remained a challenge for all developing countries. The experiences of China, India and other emerging economies show that public support, political will and concerted policy coordination are key to promoting technological capabilities over time. Greater support for education (especially at the tertiary level) and for the development of small and medium-sized enterprises, as well as financial support for larger firms and public science are important. But in addition to such domestic policy support, greater support from the international community is also needed. The *TIR 2011* proposes four mechanisms of international support. The first of these, the STI Network, was approved at the LDC IV Conference in Istanbul in May 2011.

NATIONAL POLICY FRAMEWORKS FOR RENEWABLE ENERGY TECHNOLOGIES

Integrated innovation policy frameworks at the national level are critical

The *TIR 2011* calls on national governments to adopt a new energy paradigm involving the greater use of RETs in collaboration with the private sector. Such an effort should be supported by a variety of stakeholders, including public research institutions, the private sector, users and consumers on an economy-wide basis. A policy framework that can strike an appropriate balance between economic considerations of energy efficiency and the technological imperatives of deployment of RETs in developing countries and LDCs will be the cornerstone of such an agenda for change. This will necessitate two separate but related agendas. The first should ensure the integration of RETs into national policies for climate change mitigation. The second should be the steady promotion of national innovation capabilities in the area of RETs. The latter entails addressing issues that are not only generic to the innovation policy framework, but also new issues, such as creating standards for RETs, promoting grid creation, and creating a more stable legal and political environment to encourage investments in RETs as an energy option within countries.

The *TIR 2011* proposes an integrated innovation policy framework for RETs use, adaptation, innovation and production in developing countries and LDCs. The concept of such a framework envisages linkages between two important and complementary policy regimes: national innovation systems that provide the necessary conditions for RETs development, on the one hand, and energy policies that promote the gradual integration of RETs into industrial development strategies on the other. The Report suggests that such a framework is essential for creating a virtuous cycle of interaction between RETs and science, technology and innovation.

Such a policy framework would perform five important functions, namely:

(i) Defining policy strategies and goals;

(ii) Enacting policy incentives for R&D, innovation and production of RETs;
(iii) Enacting policy incentives for developing greater technology absorptive capacity, which is needed for adaptation and use of available RETs;
(iv) Promoting domestic resource mobilization for RETs in national contexts; and
(v) Exploring newer means of improving innovation capacity in RETs, including South-South collaboration.

Policy strategies and goals are important signals of political commitment

The use and adaptation of RETs in countries requires the establishment of long-term pathways and national RE targets. These targets, although not necessarily legally binding in nature, would have to be supported by a range of policy incentives and regulatory frameworks. Defining targets is an important signal of political commitment and support, and the policy and regulatory frameworks aimed at enforcing the targets would provide legal and economic certainty for investments in RETs.

Different policy incentives for RETs innovation, production, adaptation and use are important

The successful development and deployment of any technologies, especially relatively new ones such as RETs, needs the support of several dedicated institutions responsible for their different technical, economic and commercialization aspects. Such support can be organizational (through dedicated RET organizations) or it can take the form of incentives to induce the kinds of behaviour required to meet the targets set for RETs. The *TIR 2011* lists various policy incentives for R&D, innovation and production of RETs and those that are aimed specifically at promoting technology absorptive capacity and learning related to RETs, which will be important for their wider use in national contexts. Many RETs-related policy incentives proposed have already been used by most of the industrialized countries, although developing countries are also increasingly using them or experimenting with their use. Clearly, developing countries and LDCs will need to select policy incentives that are geared to their specific situations and requirements as much as possible.

The policy incentives discussed at length in the Report pertain to two policy spheres: the innovation policy frameworks of countries and their energy policies. This is because energy policies often contain measures that have an impact on particular kinds of technologies. Ongoing reforms in the energy sectors of most developing countries offer a good opportunity to establish regulatory instruments and production obligations geared towards promoting investment in RETs and energy production based on these technologies. Policy incentives of both kinds (i.e. innovation-related and energy-related) are important to induce risk-taking by the private sector, to improve enterprise capacity to engage in learning activities, and to promote basic and secondary research in the public sector. Some of the policy incentives could be aimed specifically at the private sector, such as green economic clusters and special economic zones to boost enterprise activity, whereas others could be hybrid instruments granted to promote both public and private sector activity, such as collaborative public-private partnerships (PPPs). Yet others, such as public research grants, would be offered primarily to the public sector.

Greater domestic resources need to be mobilized for RETs

Financial incentives of various kinds can promote investment in RETs, and facilitate their quicker adaptation and utilization at the national level. These incentives need to be developed with an eye on the co-benefits of using RETs not only for electricity generation, but also more broadly as a tool for industrial development in countries. All stages of the RETs innovation and adaptation process require financing, and will depend on each country's ability to provide a mix of different kinds of financing, including venture capital, equity financing and debt financing. Particularly in developing countries that face several financial constraints on the introduction and uptake of new technologies, governments need to support the private sector in its financing of innovation activities, such as by offering loan guarantees, establishing business development banks and/or mandating supportive lending by State banks. Governments may also directly fund innovation activities through, for example, grants, low-interest loans, export credit and preferential taxation policies (e.g. R&D tax credits, capital consumption allowances).

South-South collaboration needs to be fostered

South-South collaboration presents new opportunities not only for increasing the use and deployment of RETs through trade and investment channels, but also through technology cooperation, and this can be facilitated by governments, intergovernmental organizations and/or regional development banks. Such cooperation can also be mediated by private sector owners of RETs, although this is less frequent. Technology cooperation can take several forms, ranging from training foreign nationals in the use and maintenance of RETs to supporting research in partner countries to adapt existing technologies to local needs. It can also include outright grants of RET-related IPRs or licensing on concessionary terms. The *TIR 2011* shows that in several cases developed-country institutions have been involved in bringing developing-country partners together for this sort of cooperation. The benefits of such collaboration are straightforward: it hastens the wide dissemination of RETs among developing countries along with all the commensurate benefits associated with it.

RETs can power development and a greener catch-up process

Developing countries will face different problems in RETs promotion, production and innovation, depending on their respective starting points. Nevertheless, for all developing countries, RETs present real opportunities for reducing energy poverty, and the right policies could influence the extent of benefits that could be derived from RETs use, adaptation and dissemination. This *TIR* presents five relevant findings from ongoing national and regional experiences with technology and innovation capacity-building of relevance to RETs.

First, the success of a number of emerging economies in developing technological capabilities over time is largely attributable to the role of national governments in providing strategic, concerted support for the use of RETs. However, the experiences of industrialized countries or the larger developing countries such as China and India may not be replicable in other developing countries due to their less favourable circumstances. The Report also highlights some of the policy incentives that need to be approached with caution. Of special note are those related to carbon taxes, but these may not be relevant or useful for many developing countries.

Second, developing countries should consider different kinds of energy regimes that give priority to the deployment of REs most suited to their specific contexts, while ensuring that conventional energy sources are not subsidized extensively.

Third, success in eliminating, or at least reducing, energy poverty through the use of RETs does not necessarily require large-scale projects with huge investments. Smaller initiatives have been highly successful as off-grid solutions to rural electricity, and offer considerable potential for replication.

Fourth, creating an integrated innovation policy framework of the kind outlined in this Report should not be viewed as a daunting exercise. In the developing-country context, a few incentives can go a long way towards achieving significant results. Further, many countries may already be providing several of the policy incentives discussed in the Report. The emphasis in such cases needs to be on enhanced coordination to reach targets in RETs use, promotion and innovation.

Fifth, countries will need to experiment with different policy combinations, and this learning process could have positive impacts on the co-evolution of institutional frameworks for RETs.

****

National governments in developing countries have a pivotal role to play in combining conventional sources of energy with RETs. Proactive government interventions will need the support of the international community to benefit from the full potential that RETs offer for alleviating (and eventually eliminating) energy poverty, but also simultaneously promote climate-friendly solutions on a global scale. Forging strong partnerships with the international community could also lead to the widespread dissemination of environmentally sustainable technologies worldwide, resulting in enhanced economic development and greater opportunities for large segments of populations that have been left behind in the process of globalization.

Geneva, October 2011

Supachai Panitchpakdi
Secretary-General of the UNCTAD

NOTES

1. See: http://unfccc.int/press/fact_sheets/items/4987.php.

2. Since the beginning of the eighteenth century, production and consumption patterns in the now developed countries have been dependent on energy provided successively by coal, oil and gas, and to a lesser extent by nuclear fission. The dramatic increases in the use of fossil energy (which, at current levels of annual consumption, is estimated to represent between one and two million years of accumulation) have enabled massive increases in productivity in both farming and manufacturing (Girardet and Mendoça, 2009). Such productivity growth has made possible a roughly tenfold increase in the global population over the past three centuries, accompanied by significant, if unevenly distributed, improvements in living standards.

3. Recent estimates suggest that developing countries will continue to suffer 75–80 per cent of all environmental damages caused by climate change (World Bank, 2010).

4. An innovation system is defined as a network of economic and non-economic actors and their interactions, which are critical for interactive learning and application of knowledge to the creation of new products, processes and organizational forms, among others.

5. It is estimated that electricity supply systems can easily handle up to 20 per cent of RE, and even more if systems are designed with some adjustments in intermittency.

6. Krohn, Morthorst and Awerbuch (2009) and UN/DESA (2009).

7. "Clean technologies" or "clean energies" cover a much broader range than RETs, and include clean coal, for example.

8. Broadly, the processes that fall under adaptation are those that seek to reduce/prevent the adverse impacts of ongoing and future climate change. These include actions, allocation of capital, processes and changes in the formal policy environment, as well as the establishment of informal structures, social practices and codes of conduct. Mitigation of climate change, on the other hand, seeks to prevent further global warming by reducing the sources of climate change, such as greenhouse gas (GHG) emissions.

9. The UNFCCC estimates cover only power generation, which includes carbon capture and storage (CCS), nuclear and large-scale hydro.

10. Access to environmentally sound technologies (which includes RETs) and related technology transfer has become a cornerstone of the draft UNFCCC (see Articles 4.5 and 4.7 of the draft Convention).

REFERENCES

Girardet H and Mendoça M (2009). A Renewable World: Energy, Ecology, Equality. A report for the World Future Council. Totnes, Devon, Green Books.

IEA (2000). Energy Technology Perspectives 2000. Paris, OECD/IEA.

IEA (2009). World Energy Outlook 2009. Paris, OECD/IEA.

IEA (2010). World Energy Outlook 2010. Paris, OECD/IEA.

Krohn S, Morthorst P-E and Awerbuch S (2009). The Economics of Wind Energy, European Wind Energy Association.

REN21 (2010). Renewables 2010: Global Status Report. Paris, REN21 Secretariat.

REN21 (2011). Renewables 2011: Global Status Report. Paris, REN21 Secretariat.

UN/DESA (2009). A Global Green New Deal for Climate, Energy, and Development. New York, United Nations.

UNEP and Bloomberg (2011). Global Trends in Renewable Energy Investment 2011. Analysis of Trends and Issues in the Financing of Renewable Energy. United Nations Environment Programme. Available at: www.fs-unep-centre.org/publications/global-trends-renewable-energy-investment-2011.

UNCTAD (2011). Technology and Innovation Report 2011: Powering Development with Renewable Energy Technologies. United Nations publication. New York and Geneva, United Nations.

World Bank (2010). Africa's Infrastructure: A Time for Transformation. Washington, DC, World Bank.

1 RENEWABLE ENERGY TECHNOLOGIES, ENERGY POVERTY AND CLIMATE CHANGE: AN INTRODUCTION

CHAPTER I

RENEWABLE ENERGY TECHNOLOGIES, ENERGY POVERTY AND CLIMATE CHANGE: AN INTRODUCTION

A. BACKGROUND

Sustained economic growth of the kind that leads to continuous improvement in the living standards of all people through poverty reduction rests on access to energy for all. Such a global energy access agenda requires a greater focus on energy efficiency aimed at improving ways of energy generation and use from existing resources while minimizing waste. It also requires promoting the use of other, newer or more cost-effective energy sources in all countries, which could complement the conventional energy supplies predominantly in use today. Any proposals regarding newer sources of energy need to take on board the overwhelming environmental challenge facing the world today, namely climate change mitigation.

This report focuses on the important role of technology and innovation policies in expanding the application and wider acceptance of renewable energies, particularly in the context of developing countries. It seeks to contribute to the ongoing international discourse on the need to promote the use of climate-friendly technologies globally. In the recent past, calls for reductions in emission levels of countries and proposals for low-carbon development pathways have been made internationally, particularly through the United Nations Framework Convention on Climate Change (UNFCCC).[1] At the same time, in the context of the Rio-plus-20 framework, there is an increasing advocacy for moving towards a "green economy". However, questions arise as to whether and to what extent these trends can be used to the benefit of all countries. Within the UNFCCC, polarized positions on who should shoulder responsibility for the current state of emissions and share the financial burden for mitigating climate change are based on the seemingly mutually incompatible challenges of promoting industrial development and mitigating climate change. Developing countries, in particular, face the challenge of promoting industrial development – a fundamental prerequisite for poverty reduction and equality in their societies – while reducing their reliance on conventional energy sources that have played a central role in global economic growth until recently.[2] Most of these countries also remain far more vulnerable to most of the environmental threats arising from climate change.[3]

Access to energy for all...requires the promotion of other, cost-effective energy sources.

Fostering mutually acceptable solutions to these interrelated issues has not been easy, and, as part of the United Nations' system-wide efforts in this area,[4] various international agencies have been working in many different socio-economic policy domains, including trade, industrial, investment, and technology and innovation policies. The United Nations Secretary-General's high-level Advisory Group on Energy and Climate Change (AGECC) has identified several goals with the aim of achieving universal access to energy and reducing global energy intensity by 40 per cent by 2030 (AGECC, 2010). UNCTAD's own work and policy advice has been addressing the challenges posed by climate change to growth and development in various ways (see, for example, UNCTAD 2010a and 2010b). Building further on the work in the United Nations system and within UNCTAD, this *Technology and Innovation Report (TIR)*

Developing countries face the challenge of promoting industrial development while reducing their reliance on conventional energy sources.

2011 focuses on policy issues and options to address energy poverty and climate change mitigation through the greater use of renewable energy technologies (RETs).

This *TIR* argues that a mutually compatible response to the dual challenge of reducing energy poverty and mitigating climate change requires a new energy paradigm. Such a paradigm would have RETs complementing (and eventually substituting) conventional energy sources in efforts to alleviate energy poverty across the developing world. This is a realistic paradigm given that the world is faced with energy poverty issues that cannot be resolved using conventional fuel sources without risking irreversible climate change. Moreover, rapid technological developments in RETs not only rendered them cheaper than they were a decade ago, but also the technological characteristics of many established RETs today enable them to be more easily combined in complementary ways with conventional energy sources. This makes it easier to envisage energy solutions that mix renewables with conventional energy in the short term or mid-term, with the view to ultimately replacing conventional energy in the long term in the interest of climate change (UN/DESA, 2011). In finding newer energy solutions that integrate renewable energy with existing energy sources, developing countries will need to develop technology and innovation capabilities. This will be necessary not only to enable the greater dissemination, adaptation and use of existing RETs, but also for promoting newer technological changes in renewable energy that will be important for a sustainable future.

A mutually compatible response to the dual challenge of reducing energy poverty and mitigating climate change requires a new energy paradigm...

...with energy solutions that mix renewables with conventional energy in the short and mid-term, ultimately replacing conventional energy in the long term.

B. A NEW URGENCY FOR RENEWABLE ENERGIES

Four current trends lend a new urgency to the need to explore how far and how easily RETs could serve energy needs worldwide. First, ensuring universal access to conventional energy sources using grids entails high costs, which means that developing countries are unlikely to be able to afford the costs of linking new households, especially those in rural areas, to existing grids.[5] Second, the climate change debate has injected a greater sense of urgency into searching for newer energy options, as a result of both ongoing policy negotiations (i.e. the impending negotiations under the aegis of the UNFCCC) and the greater incidence of environmental catastrophes worldwide.[6] This makes it imperative for countries to reach some level of consensus on mitigating ongoing climate change effects as soon as possible. Third, from a development perspective, the recent financial and environmental crises have caused major setbacks in a large number of developing countries and least developed countries (LDCs), resulting in their further marginalization from the global economy. The LDCs and many developing countries suffer from severe structural vulnerabilities that are a result of their patterns of integration into the global economy (UNCTAD, 2010b). Promoting low-carbon, climate-friendly development while fostering inclusive economic growth in these economies is an urgent imperative for the international community. Lastly, there are severe inequalities within developing countries themselves, and lack of access to energy affects the poorest of the poor worldwide, impeding their ability to enjoy the basic amenities of modern life that are available to others at the same level of development.

1. An energy perspective

The energy revolution that served as a major impetus to industrial development can be traced back to the introduction of steam as a source of energy, which was later followed by the discovery of oil and gas. Use of these energy resources has enabled steady economic growth globally, contributing to a 2 per cent annual increase in industrial production. Over the decades, dramatic increases in the use of energy from fossil fuels have enabled unprecedented productivity growth, accompanied by significant, albeit unevenly distributed, improvements in living standards. Thus, countries that have achieved high levels of development also present higher levels of energy use per capita and per unit of output than countries at lower levels of development (Martinez and Ebenhack, 2008).[7] At the more advanced stages of development, economies show a decline in the energy intensity of output because of structural change towards less energy-intensive service activities and more widespread availability of more efficient technologies. Nevertheless, energy use continues to grow in the industrialized economies, and indeed, very significant increases in energy demand have been forecast for the developing world (see boxes 1.1 and 1.2 below).[8]

Very significant increases in energy demand have been forecast for the developing world.

Box 1.1: Energy demand and the role of RETs

Under the International Energy Agency's (IEA) New Policies Scenario laid out in its *World Energy Outlook 2010*, achieving basic universal access to energy by 2030 would require an additional 950 terawatt-hours (TWh) of electricity generation and would mean an additional generating capacity of 250 gigawatts (GW). A mix of different RETs, including extensive use of off-grid and mini-grid applications, will be needed (380TWh on-grid, 400TWh via mini-grids and 172TWh via off-grid applications). Developing countries would require most of this additional electricity generation because, under the scenario, energy poverty will remain more or less a developing-country problem by 2030. The main problem regions are sub-Saharan Africa (which would require an additional 462TWh), India (requiring 245TWh) and other parts of Asia (requiring 221TWh).

Source: UNCTAD, based on IEA (2010).

Box 1.2: Africa's energy challenge

With 5 per cent of global primary energy use and 15 per cent of the world population, per capita energy consumption in Africa is only a third of the global average. Nearly half of the current energy use is traditional biomass, a major cause of health problems and deforestation. In 2009, 657 million Africans relied on traditional biomass and 587 million people lacked access to electricity. Limited and unreliable energy access is a major impediment for economic growth. In the coming decades the energy mix will have to change to modern fuels, the per capita energy use will increase and the population will grow much faster than the global average. Together these three factors will put tremendous pressure on future African energy supply.

Energy access is an important issue directly related to income and poverty. Access to modern energy rises from virtually zero for the lowest income quintile to 70-90 per cent for the highest income quintile (Monari, 2011). Access can be split into two types: access to electricity for residential and commercial use and access to modern cooking fuels.

Adequate electricity provision is a challenge for industry and policy makers. Between 1990 and 2005, the poor performance of the power infrastructure retarded growth, shaving 0.11 per cent from per capita growth for Africa as a whole and as much as 0.2 per cent for Southern Africa (Foster and Briceno-Garmendia, 2010). In sub Saharan Africa, 30 out of 48 countries experience daily power outages. These cost more than 5 per cent of gross domestic product (GDP) in Malawi, Uganda and South Africa, and 1-5 per cent in Senegal, Kenya and Tanzania (Foster and Briceno-Garmendia, 2010). Diesel generators are used to overcome outages and more than 50 per cent of power generation capacity in countries such as the Democratic Republic of Congo, Equatorial Guinea and Mauritania and 17 per cent in West Africa is based on diesel fuel. The resulting generation cost can easily run to $400 per megawatt-hour (MWh). Reliable, affordable, low cost power supply is needed for economic growth. Renewable energy can play an important role in filling this gap.

The International Renewable Energy Agency (IRENA) estimates that Africa spends about $10 billion per year on the power sector: $2.27 billion for grid extension, $4.59 billion for grid supply, $1.37 billion for off-grid renewable electricity, $1.07 billion for policy/regulation and $0.76 billion for efficient use of electricity (Monari, 2011). What would be needed is an investment of $40.6 billion per year, consisting of $26.6 capital expenditure and $14.0 billion operation and maintenance. This implies a quadrupling of investments. Annual capacity additions would need to rise to 7 GW per year. The most remarkable feature of African energy systems is the fact that the continent exports 40 per cent of the energy it produces. This is largely oil and gas that is exported from the North and West African countries. As such, energy scarcity is not an issue for Africa as a whole. The problem is the uneven distribution of the resource and the fact that the indigenous population is too poor to afford commercial fossil energy.

Source: IRENA (2011), forthcoming.

Energy consumption can have a variety of impacts on productivity, depending on the level of development of countries. In countries at more advanced levels of industrialization, increased availability of high-quality[9] energy generally allows greater use of advanced machinery and transport equipment, which raises labour productivity. Better quality energy supply also allows a reduction in the amount of capital needed to ensure back-up capacity (e.g. individual generators).[10] Improving the reliability of energy supplies for electricity for lighting and for the operation of information and telecommunications equipment in developing countries is therefore expected to have an immense positive impact on the quality of life. Access to energy will also help promote the implementation of several Millennium Development Goals (MDGs), especially those relating to education and health, with commensurate positive implications for the greater availability of human resources for productive activities.[11]

Energy consumption can have a variety of impacts on productivity, depending on the level of development of countries.

2. A climate change perspective

Use of conventional energy sources (primarily fossil fuels) are believed to have led to a rise in GHG emissions and to a resulting increase in global average temperatures since the mid-twentieth century (IPCC, 2008). The fundamental conclusions of the most recent assessment report of the IPCC are that climate change is the result of human activity, that the ongoing rate of climate change will have devastating effects if left unchecked, and that the costs of action for mitigation and adaptation would be significantly lower than the costs of inaction.[12] Along the same lines, the Stern Review on the Economics of Climate Change[13] has estimated that the cost of climate change would amount to a loss of at least 5 per cent of global GDP per annum, and could even reach 20 per cent, while actions to counter the worst effects of climate change could cost about 1 per cent of global GDP (2 per cent in more recent updates) (Stern, 2007). It has also been argued that the effects of climate change, if left unchecked, could become a threat to global peace and security.[14]

The ongoing rate of climate change will have devastating effects if left unchecked.

Dubbing climate change as a global market failure, these reports present various proposals for emission reductions (discussed in chapter IV of this *TIR*). The contentious issue here is the perceived divide between the interests and obligations of developed and developing countries. The latter believe that developed countries—the source of most of the past and current emissions of GHGs — should act first and bear most of the costs of reducing GHG emissions. The varying levels of historical responsibility of different countries for the climate change problem, as well as the extreme differences in the financial capacities of countries have also led to discussions at the global level on who should bear the major costs of climate change mitigation efforts. Additionally, mechanisms and incentives for greater private sector involvement – including technology transfer through the Clean Development Mechanism (CDM), carbon credits and tradable emission certificates – have all proven to be rough terrain in international negotiations.

Nevertheless, these debates have given a much-needed impetus to international discussions on RETs and how they could help to resolve the dual needs of reducing energy poverty and mitigating climate change. Several discussions on how to make relevant technologies and finances available for RETs have been taking place in the international debates on climate change. At the same time, the development of green businesses and the concept of the green economy have both emerged as possible effective responses for mitigating climate change.

3. A developmental perspective

The extent to which energy policies can accommodate new incentives and mechanisms to promote low-carbon growth trajectories will be different for each country depending on its stage of development. Industrialized countries maintain high living standards and consumption patterns that have been dependent on high absolute and per capita levels of carbon dioxide (CO_2) emissions. They present the largest potential for quick reductions of carbon emissions through changes in consumption patterns.[15] Industrialized countries could potentially improve the technology mix of their energy generation policies by making RETs more widely available in their countries and with relatively greater ease. These changes are already being witnessed in several European economies, such as Denmark, Germany, Spain, and the United Kingdom.

Larger developing countries, such as China and India, could also benefit from gradual efforts to reduce the carbon intensity of their economies, especially as they push ahead with industrial development over the next decade. RETs have a clear role to play in this development. A priority in this regard will be to identify strategies to weaken the association between the increase in GDP per capita and carbon emissions.

The LDCs present levels of energy intensity of output that are close to the world average, although their GDP per capita is around seven times lower than the world average. Given that they are particularly affected by energy poverty (see section B.4 below) and that they still produce low levels of GHG emissions, along with the fact that they have not contributed in any significant way to the historical build-up of GHG concentrations in the atmosphere, the main contribution of the LDCs to the rebalancing of the world's energy system should be as beneficiaries, that is, through the provision of modern energy services to those that currently lack them. This should be done in a way that relies as much as possible on RETs or other low-carbon-intensive technologies. Although this may not be practicable or cheaply available in every case, it is important that all developing countries including LDCs, embark on a transition to a low-carbon economy as soon as possible. In the absence of this, their future growth strategies will get locked into high-carbon technologies that must become obsolete in the short to medium term if climate change on a catastrophic scale is to be avoided. An example of this is China, whose own industrial development has been enabled by large investments into coal plants made some decades ago. Despite China's extensive shift towards RETs, the coal plants will take some more decades to become obsolete.

All developing countries including LDCs, should embark on a transition to a low-carbon economy as soon as possible.

4. An equity and inclusiveness perspective

Substituting or complementing conventional energy sources with RETs in order to promote greater access to energy raises all the issues that are currently prevalent in the context of energy poverty and development. Will such a new energy paradigm that envisages a greater role for RETs be able to create more employment? Will it be applicable in remote rural areas which are hard to connect to the conventional energy grid? Will it be applicable and easy to use by individual users, but at the same time have the potential for scale-up within enterprises, firms and sectors? Would it alleviate, at least partially, the difficulties faced by vulnerable social groups affected by poverty (e.g. rural populations, women, children and indigenous groups) so that they can devote more time and attention to income-generating activities?

A significant aspect of renewable energy use is the possibility of devising semi-grid or off-grid rural installations that promote greater access to energy in developing countries than that provided by conventional energy sources which rely extensively on grid connections. This flexibility enables the better consideration of demand-side requirements in designing renewable energy solutions. For instance, the solar supply heating systems or solar lamps that can be used in rural areas for electrification can improve the quality of life in contexts where on-grid solutions are currently not possible.[16] Of the 1.4 billion people not connected to electricity grids globally, approximately 85 per cent live in rural areas where technologies such as solar pumps, solar photovoltaic installations,

Of the 1.4 billion people not connected to electricity grids globally, approximately 85 per cent live in rural areas.

small wind, mini-hydro and biomass mini-grids offer high potential and cost advantages over traditional grid extension (IEA, 2010, chapter 8). As a result, when pitted against the current state of underdeveloped energy infrastructure in developing countries, RETs could help to reduce energy poverty in many novel ways, and at the same time also reduce social inequalities through the creation of new jobs in the application processes of RETs. Therefore, national strategies for RET development, production, adaptation and use in developing countries need to be well integrated into policies for industrial development and poverty reduction.

When pitted against the underdeveloped energy infrastructure in developing countries, RETs could help to reduce energy poverty in novel ways.

From an equity perspective, subsidies have had a significant distorting effect on conventional fuels versus RETs and biofuels. Developing countries still allocate a significant amount of their financial resources to subsidize conventional fuels. In 2009, subsidies amounting to $312 billion were spent on fossil fuel energy worldwide, but mainly by developing countries,[17] compared with $57 billion spent worldwide on subsidies for RETs and biofuels (IEA, 2010).[18] It is estimated that a gradual phase-out of these subsidies between 2013 and 2020 could reduce global primary energy demand by 5 per cent, oil demand by 4.7 million barrels/day and CO_2 emissions by 5.8 per cent by 2020 (IEA, 2009 and 2010). While the distributional effects of this reduction in fossil fuel subsidies need to be fully analysed, it is generally acknowledged that in most countries it is the middle and higher income groups that benefit the most from fossil fuel subsidies. Therefore, a gradual transfer of subsidies from fossil fuels to RETs, particularly if these are applied to reducing energy poverty, is likely to improve both equity and efficiency. A phasing out of subsidies in ways that target the middle and higher income groups in all countries, while protecting the lower income groups, could also be desirable depending on the situation in each country. These options and the accompanying issues are discussed in greater detail in chapter V of this Report.

The overwhelming policy consideration for developing countries is whether such an agenda offers the hope of an adequate energy supply...to jump-start industrial development.

C. ENERGY POVERTY AND GREENER CATCH-UP: THE ROLE OF TECHNOLOGY AND INNOVATION POLICIES

In much of the industrialized world, issues relating to climate change have begun to revolve around the notion of the "green economy". Still very much an evolving concept, the green economy can be defined as economic development that is cognizant of environmental and equity considerations and promotes the earth's environment while contributing to poverty alleviation. As a recent report by a Panel of Experts to the United Nations Conference on Sustainable Development (UNCSD) notes, the concept has gained currency in the light of the recent multiple crises that the world has seen (climate, food and financial) as a means to promote economic development in ways that "...will entail moving away from the system that allowed, and at times generated, these crises to a system that proactively addresses and prevents them" (UN/DESA, UNEP and UNCTAD, 2010: 3). How far this can actually be made to happen in an inclusive way is still much debated. The "green economy" and "clean energies" agenda are very appealing to most developed countries but are viewed with skepticism and concern by developing countries. The overwhelming policy consideration for developing countries is whether such an agenda offers the hope of an adequate energy supply at reasonable costs to jump-start industrial development and structural change, while at the same time promoting the shift to a low-carbon, sustainable development path. They are also concerned about the potential use of the green agenda as an instrument of trade protectionism. To ease the lingering concern, the transition of developing countries to the green economy must be supported through finance and investment, technology transfer and other supportive measures (see Chapter IV). Issues of technological change and innovation capacity therefore need to be at the forefront of this discourse and this *TIR* seeks to contribute to new policy insights in this extremely complex area. In the absence of such a focus, the transition to the green economy and strategies for sustainable development which seek to promote greater use of RETs are likely to be constantly undermined by the lack of technological and innovation capabilities, which are required not only for research and development (R&D) and innovation of new RETs, but also for adaptation, dissemination and use of RETs.

1. Towards technological leapfrogging

Only a limited number of developing countries (e.g. Brazil, China and India) are steadily making their mark as developers of RETs and their firms are gaining significant markets in renewables globally (as discussed in chapter III). Some studies and authors have also noted that expertise in developing countries has been concentrated to a large extent in less technology-intensive RETs such as biofuels, solar thermal and geothermal. Many of these countries either have existing expertise, or stand good chances of developing such expertise and of becoming competitive exporters of such technologies. Furthermore, in the case of China and India, the sizeable domestic markets have been springboards for export success, driven, as in the member countries of the Organization for Economic Co-operation and Development (OECD), by ambitious domestic targets for renewable energy generation. For instance, China installed 16.5 GW of domestic wind power capacity in 2010 – more than any other country and more than three times the amount installed in the United States (Ernst & Young, 2011). India ranked third with a capacity addition of 2.1 GW (Balanchandar, 2011).

The obvious question for other developing countries, and for the global community as a whole, is whether the capabilities in renewable energy technologies demonstrated by the BRICS (Brazil, the Russian Federation, India, China and South Africa) represent special cases. To some extent they do: the prerequisites for competitive production of many RETs are a workforce with advanced technical training, supporting industries and services in the high-tech areas, access to finance, ample government assistance and a large domestic market, all of which would seem to favour larger emerging developing countries over smaller, poorer developing countries and LDCs. In all developing countries, promoting technological learning and innovation has remained a challenge historically. The successes of China, India and other emerging economies shows that public support, political will and concerted policy coordination are key to promoting technological capabilities over time. Greater support for education (especially tertiary education) and for the development of small and medium-sized enterprises, and financial support for larger firms as well as public sector research are all important. In the case of RETs too, the most relevant lesson from both China and India is the importance of constant policy support by governments for the promotion of RETs. However, there are other factors that also need to be considered when extrapolating from the more advanced developing countries. China, for example, may be heavily investing in RETs, but it has already experienced significant economic growth and industrial development through investment in conventional energy, which explains much of its global economic competitiveness today.

Lastly, most RETs are still developed and held by industrialized countries. As a result, there is a tendency for firms in developing countries, which are largely technology followers in this field, to underinvest or they have difficulties in accessing technologies and related know-how from abroad and in learning how to use it effectively. Most proponents of the leapfrogging argument tend to argue that since technologies are already available, they can be used at marginal costs by developing countries and LDCs to simply circumvent being "locked into" the conventional, resource-intensive patterns of energy development. Leapfrogging is also possible, it is claimed, because RETs can contribute to building new, long-term infrastructure, such as transport and build-

Only a limited number of developing countries are steadily making their mark as developers of RETs.

The most promising way to promote leapfrogging through RETs would be to integrate them holistically as part of the technology and innovation policy framework.

ings, in ways that promote cogeneration of technologies (Holm, 2005).[19] This Report suggests that the most promising way to promote leapfrogging through RETs would be to integrate them holistically as part of the technology and innovation policy framework of countries.

2. The crucial role of technology and innovation policies

Developing countries will need to strengthen their innovation systems...
... to enable wider RET dissemination and to promote a greener catch-up process.

Technology and innovation policies can promote and facilitate the development, acquisition, adaptation, deployment and use of RETs to support sustainable development and poverty reduction in developing countries and LDCs. Although many of the RETs needed in order to meet a larger share of the global energy demand already exist, or are on the verge of commercialization (IPCC, 2008), the knowledge and technological capabilities required for their transfer to developing countries and LDCs are not easily accessible. The costs and possibilities of making these technologies available and adapting them to local contexts in developing countries and LDCs are also unclear. Developing countries will need to strengthen their innovation systems[20] through innovation policy frameworks that foster capacity and linkages to enable wider RET dissemination and to promote a greener catch-up process. International support to developing countries through various channels will be essential for this effort, including financial support and North–South, South–South and triangular cooperation, and effective technology transfer mechanisms. All of these will be necessary complements to the development of local capacities for RETs.

Energy security and technological capabilities have a virtuous relationship.

The advantages of using RETs will not accrue automatically. The untapped opportunities offered by already developed technologies and the unprecedented amount of information and knowledge are neither directly nor easily available. Not only are strong domestic technology and innovation policies needed, but also greater international support is required to make the international trade and intellectual property regime more supportive of the technological needs of developing countries and LDCs. Promoting greater access to RETs and support for use and adaptation of these technologies through all means possible will be important for developing countries to sustainably integrate these processes into efforts aimed at capital formation and transformation of their productive structures.

This *TIR* identifies five distinct issues that stand out in the debates on technology and innovation in RETs that are of particular relevance to developing countries and LDCs. First, structural transformation that supports the economic development of countries relies strongly on the growth of national technological capabilities. Wider dissemination and use of RETs can be a valuable part of their overall industrialization effort. The lack of energy is a constraint that applies not only to the manufacturing sector, which in most low-income countries is nascent, but also to other sectors that are potentially important to the process of industrialization and development, such as services, tourism and agricultural processing, which depend on reliable, high-quality power supply. It is therefore important to recognize that energy security and technological capabilities have a virtuous relationship: energy security is a key aspect of the physical infrastructure that promotes enterprise growth in the early stages of structural change, and technological capabilities are a fundamental prerequisite for greater adaptation and use of RETs within domestic economies.

Second, incoherent, and often conflicting, policy developments at the multilateral level tend to adversely affect national aspirations for technological empowerment in developing countries in this highly complex terrain (see chapter IV). Although climate change will affect all countries and communities worldwide, developing countries (especially LDCs in Africa and South Asia) will shoulder a disproportionate burden from the fallout resulting from climate change, including increasing climatic variations, extreme weather events and natural disasters. The ongoing debates on climate change reflect the diverse positions of countries on how the burden should be shouldered.

Third, the issue of greater transfer of climate-friendly technologies that has been a key element in the global debate on climate change is intricately linked to technology and innovation infrastructures in countries. The UNFCCC has repeatedly called on developed countries to take steps to promote the transfer of technology to developing countries, and technology issues will remain a key component of the Conference of the Parties' work within the framework of the UNFCCC for years to come. Noting this, the Bali Action Plan called for greater attention to "technology development and transfer to support action on mitigation and adaptation", including the consideration of "effective mechanisms and enhanced means for the removal of obstacles to, and provision of financial and other incentives for, scaling up of the development and transfer of technology to developing country Parties in order to promote access to affordable environmentally sound technologies".[21] In the renewable energy sector, recent evidence shows that basic approaches to solving technological problems have long been off-patent, and therefore can be adapted and disseminated in developing countries *provided that* some technological prerequisites are met. This points to the need for greater attention to strengthening the technological absorptive capacity of countries through coordinated policy support, in addition to making existing technologies available and aiding in their greater diffusion.

Fourth, RETs will remain a distant goal as long as they are prohibitively expensive. Governments need to intervene through the design of appropriate regulations and innovation policies to promote public and private financial investment in RETs, and to ensure the wider use of RETs across all productive sectors of the economy. Innovation in RETs is moving at a fast pace globally, but left on its own, or left to the "market", it is unclear to what extent this pace will continue globally and to what extent it will lower the prices of these technologies for use at the individual household and firm level in the medium term. Governments in developing countries will need to encourage a broader focus on RETs that ranges from use, to adaptation, to production and innovation, in collaboration with the private sector and users.

Finally, RETs form part of the wider debate on emerging patterns of investment and technology that fall under the umbrella of the green economy. At a fundamental level, the concept of the green economy itself has been highly contested. Some argue that calling for large-scale investments in developing countries to facilitate the transition to green economy imposes uneven costs, thereby creating an additional burden on already disadvantaged groups of people. The challenge is to ensure that the green economy concept, which will also be the focus of the Rio-Plus-20 Framework, is structured in a way that it does not adversely affect ongoing productive activities in developing countries, while helping their transition to "green" modes of development. Numerous issues will need to be addressed in this context, including patterns of trade, technological upgrading and specialization.

Governments in developing countries need to encourage a broader focus on RETs that ranges from use, to adaptation, to production and innovation.

Analyzing these five issues at length, this Report argues that there are numerous benefits of RETs for developing countries. The potential impacts of RETs in terms of reducing energy poverty, generating employment and creating new production and innovative activity add to their environmental advantages. Several established RETs have significant potential to contribute to a broad range of development goals. It is beyond the scope of this Report to address the whole range of policy implications of all RETs in the very different contexts of the various categories of developing countries. It therefore focuses on those that are (a) already mature enough to make practical contributions to policy objectives in the short term, but are sufficiently recent in their commercialization to present challenges with which policymakers may be less familiar, and (b) particularly appropriate to the objective of reducing and eventually eliminating energy poverty in developing countries as complements (and eventually substitutes) to conventional energy sources. The two subsections below define the key terms and present the structure of this *TIR*.

3. Definitions of key terms

Very significant gains in terms of health, education, gender equality and income generation could be expected from the provision of basic electricity.

Two important terms that need to be explained clearly at the outset are energy poverty and renewable energy technologies. These terms are discussed below, based on widely accepted definitions of the concepts.

a. Energy poverty

According to a commonly used definition, energy poverty implies lack of access to modern energy services, which includes lack of household access to electricity and clean cooking facilities (i.e. clean cooking fuels and stoves, advanced biomass cooking stoves and biogas systems) (see AGECC, 2010; IEA, 2010). It has been estimated that access to 100 kWh of electricity and 100 kilograms of oil equivalent (kgoe)[22] of modern fuels per person/year represent the minimum level defining energy poverty (IEA, 2010). By implication, anything below this level would amount to energy poverty. Other criteria for defining of energy poverty relate to the extent of availability of electrical and mechanical power for income-generating activities, supply reliability (for households as well as for enterprises) and affordability.

The truly transformative effects of modern forms of energy only manifest themselves when energy can be applied to economic activity on a significant scale.

In its report, the AGECC (2010) defines its proposed goal of achieving universal energy access as "access to clean, reliable and affordable energy services for cooking and heating, lighting, communications and productive uses". This definition goes beyond the basic human needs that would be covered by the IEA's minimum threshold of 100 kWh plus 100 kgoe of modern fuels; it also includes access to electricity, modern fuels and other energy services to improve productivity in areas such as agriculture, small-scale industry and transport. It is this broader definition of energy poverty that is adopted and in this *TIR*, along with a discussion of the related issues.

The rationale for this choice is not based on the view that the benefits of ending energy poverty in its most restricted definition would be modest. On the contrary, as argued earlier, very significant gains in terms of health, education, gender equality and income generation could be expected from the provision of basic electricity for lighting, for the use of information and communication technologies (ICT), for health care, and for cooking. However, the truly transformative effects of the availability of modern forms of energy only manifest themselves when energy can be applied to economic activity on a significant scale so that it contributes to improving livelihoods in such a way as to change economic structures and relationships, even if only at the local level. Access to energy has long-term effects when it has a direct impact on livelihoods and revenue generation in addition to improving living standards. Such impacts can be ensured by enhancing the productivity of an existing production process or by enabling new lines of activity that will generate employment and local demand. This can happen by freeing labour from subsistence activities so that it can be employed in higher value-added ones which generate surplus that can be saved and invested, or by enabling the operation of even small industries for serving local markets, usually beginning with the transformation of agricultural products. It is only when energy services enable larger scale economic undertakings and greater cooperation between economic actors, as well as broadening the reach and hence the efficiency of markets that they become drivers of long-term development.[23]

b. Renewable energy technologies

RETs are diverse technologies that convert renewable energy (RE) sources into usable energy in the form of electricity, heat and fuel. And because some of them can be deployed for many different applications, they can play a significant role in diverse situations. Simply put, renewable energy refers to energy generated from naturally replenishable energy sources (box 2.1 of chapter II). The main types of RETs include hydropower, bioenergy (biomass and biofuels), solar, wind, geothermal and ocean energy. Currently, the so-called second generation RETs, including solar energy in its various forms (photovoltaic, heating and thermal or concentrated), wind power technologies and several modern forms of biomass use technologies (particularly biogas digesters), are the ones that are registering the fastest deployment growth rates in both developed and developing countries, and their upfront costs are declining fast (REN 21, 2010).

These technologies can be applied in a broad range of development contexts and, in particular, demonstrate significant potential for application in rural as well as urban areas in developing countries through small-grid and non-grid systems (ESMAP, 2007; REN 21, 2010). Accordingly, most of the information and discussion in this Report is presented mainly from the perspective of the implications of wind, solar and modern biomass RETs for development, although this does not preclude consideration of other forms of RETs in specific contexts in developing countries. There are social costs and consequences associated with some RETs such as large hydro and biofuels (see chapter II). The Report recognizes that countries are faced with important trade-offs when making development choices. Some of these trade-offs may be very complex, and require consideration of how best to address them in specific national socio-cultural contexts.

D. ORGANIZATION OF THE REPORT

Following this introduction, chapter II describes current technological trends in renewable energies, tracing trends in development and use across a broad range of RETs. The chapter examines ways in which RETs could potentially complement traditional sources of energy in developing countries based on their varied technological characteristics. It also describes the declining costs of use of some RETs, and highlights the technological progress that makes them more cost competitive. Using several examples from developing as well as developed countries, the case for broader applicability of such technologies is presented.

Chapter III presents the framework for technology and innovation in the context of RETs. The presence or absence of the elements of the framework will determine the ability of developing countries to harness the potential of RETs as an engine of sustainable development. It presents the mutually dependent relationship between countries' technology and innovation capacity and the wider dissemination and use of RETs, and analyses the role of interdependent factors. Chapter III argues that there is a need for greater policy intervention and support within countries as part of their innovation policy frameworks to promote the innovation, production, use and diffusion of RETs, thereby harnessing energy solutions for sustainable development processes. Such deliberate policy actions taken in technology and innovation policy frameworks will help to: (a) integrate RETs within the socioeconomic development strategies of countries; and (b) provide the requisite national parameters that are necessary to foster technological absorption capacity. These actions will increase the demand for RETs in developing countries and LDCs creating the requisite economies of scale in use and diffusion that are required at the global level to drive down the prices of these technologies. Apart from a reduction of energy poverty, the chapter argues for a need to clearly integrate use of RETs into strategies for poverty reduction and job creation, especially for the more economically vulnerable groups in developing countries and LDCs.[24]

The main types of RETs include hydropower, bioenergy (biomass and biofuels), solar, wind, geothermal and ocean energy.

Chapter IV analyses four important policy challenges related to climate change and renewable energy technologies in the international policy context. These are: (i) the need for a new international narrative that focuses on energy, (ii) financial support for RETs within the international architecture on

RETs can be applied in a broad range of development contexts and, in particular, have significant potential for application in rural areas.

climate change, (iii) technology transfer, and (iv) intellectual property rights (IPRs). These issues have been, and remain, central to all debates and decisions of the UNFCCC and the Kyoto Protocol. Many of these discussions refer to environmentally sustainable technologies or clean technologies,[25] of which RETs form a subset. Developing countries will need greater international support to promote technology and innovation capacity for RETs, which needs to be factored in as an urgent priority in the international negotiations and developments. Noting the limitations of the ongoing international negotiations to deal with the important issue of promoting RETs, the chapter stresses the need for a new international approach to energy that factors in technological issues related to RETs more robustly in the climate change negotiations and the Rio-Plus-20 framework. It makes concrete suggestions on how the international policy framework could support the use of RETs through financing, technology transfer and favourable treatment of IPR issues. Each of these issues are examined in terms of key international developments and the main hurdles that remain to be overcome in order to ensure that the international discourse on these issues serves the needs of science, technology and innovation (STI) for RETs development in developing countries.

Chapter V presents elements of a national integrated innovation policy framework for RETs to promote simultaneously the diffusion and use of RETs, as well as their production and innovation, as applicable in different developing-country contexts. It considers ways of mobilizing much-needed investment, and the roles of public and private finance in meeting those needs. Many of the policy incentives discussed in this chapter have been used more widely in the industrialized countries, and there has been an increasing level of use and experimentation in developing countries. With this in mind, the analysis seeks to focus the discussion on the developing-country context as much as possible.

NOTES

1 The UNFCCC was conceived at the United Nations Conference on Environment and Development in 1992. The Convention aims to reduce greenhouse gases (GHGs) in an effort to mitigate climate-change-related effects on the earth's atmosphere. UNFCCC is also the name of the United Nations secretariat that is in charge of implementing the treaty and the negotiations related to it.

2 Since the beginning of the eighteenth century, production and consumption patterns in the more developed countries have been dependent on energy provided successively by coal, oil and gas, and to a lesser extent by nuclear fission. The dramatic increases in the use of fossil energy (which, at current levels of annual consumption, is estimated to represent between one and two million years of accumulation) have enabled massive increases in productivity in both farming and manufacturing (Girardet and Mendoça, 2009). Such productivity growth has made possible a roughly tenfold increase in global population over the past three centuries, accompanied by significant, if unevenly distributed, improvements in living standards.

3 Recent estimates suggest that developing countries will continue to bear 75–80 per cent of all environmental damages caused by climate change (World Bank, 2010).

4 Coherence in this area within the United Nations system is ensured through UN-Energy, which was established as part of the follow-up to the World Summit on Sustainable Development (WSSD). UN-Energy is concerned with policy development in the energy area, and its implementation. It also maintains a database of major ongoing initiatives throughout the system based on the UN-Energy work programme at global, regional, sub-regional and national levels. The Johannesburg Plan of Implementation (JPOI), decisions taken at CSD-9, Agenda 21 and the Programme for Further Implementation of Agenda 21 serve as the basis for action on energy (see http://esa.un.org/un-energy/index.htm).

5 It is estimated that connecting each family unit will cost roughly $2,000.

6 The Intergovernmental Panel on Climate Change (IPCC, 2008) has provided estimates of increasing climatic risks and catastrophes on a global scale as a result of climate change. A more recent report by the World Bank (2010) notes that new climatic risks in hitherto unknown places are becoming common. For example, floods, once rare in Africa, are now becoming common, and the first hurricane ever recorded in the South Atlantic hit Brazil in 2004.

7 While there is a clearly established relationship between economic growth and energy consumption, the direction of causation remains controversial. Efforts to establish it by empirically employing Granger or Sims techniques offer mixed results and therefore ambiguous policy implications (see, for example, Payne, 2010). Others believe that, like good health, energy use is a contributor to, as well as a consequence of, higher incomes. Conversely, energy poverty is a cause as well as a consequence of income poverty (Birol, 2007).

8 For example, IEA (2010) forecasts that world energy consumption will increase by 49 per cent in 2035, compared with the consumption rate in 2007 (from 495 quadrillion British thermal units (Btu) in 2007 to 739 quadrillion Btu in 2035). It also estimates that non-OECD economies will consume 32 per cent more energy than OECD economies in 2020 and 63 per cent more in 2035 respectively.

9 This refers to energy from sources that provide a steady supply of energy in a controlled, safe and stable manner, such as coal, oil and gas (either used directly or through the generation of electricity). These can be obtained through technically simple and low-cost processes, besides being portable and having a high energetic content (capacity to do work) per unit of mass.

10 In a study on African infrastructure, the World Bank (2009) estimates that the losses imputable to poor quality energy supply can be as much as 2 per cent of potential growth per year as a result of outages, excessive investment in back-up capacity, energy losses and inefficient use of scarce resources.

11 Nordhaus (1994) provides a striking illustration, viewing the cost of an hour's evening reading time in terms of the average time of work that would buy the necessary means of lighting. In ancient Babylon, it took the average worker more than 50 hours to pay for that light from a sesame oil lamp. In the United Kingdom in 1800, more than six hours of work were still needed to pay for an hour's worth of a tallow candle. Today, in advanced economies, electricity and compact fluorescent bulbs have lowered the cost to less than a second.

12 See: http://unfccc.int/press/fact_sheets/items/4987.php.

13 Commissioned by the Government of the United Kingdom.

14 See, for example, statements by representatives of several Member States of the United Nations and by Secretary-General Ban Ki-moon at the debate of the Security Council of the United Nations on 17 April 2007. (DPI'S PRESS RELEASE SC/9000 OF 17 April 2007, available at http://www.un.org/News/Press/docs/2007/sc9000.doc.htm).

15 Since the populations of developed countries are also the ones that are the least vulnerable to the consequences of climate change, modifications in their consumption patterns need to be articulated in way that is acceptable to the general electoral public in these countries. A number of interesting proposals have been made in this regard. One is the "2000 watt (W) society" initiative of the Swiss Federal Polytechnic School in Zurich, backed by the Swiss Federal Office of Energy. The proposal includes changes that would cut the average per capita energy use in the developed world to 2000 watts (17,520 kilowatt-hours (kWh)) by 2050 (or 2030 in the version proposed by the Swiss Solar Society). This is roughly equivalent to the current world average for energy use, and was the level of use of a Swiss citizen in the 1960s (corresponding to one of the world's most affluent societies at the time). It is also about one third of the current average energy use in Western Europe or one sixth of that of the United States. The proposal emphasizes the need for technological innovation in RETs and materials, and investment in and renovation of housing and other infrastructure, particularly transport. The "2000 watt society" could thus be achieved without

compromising the levels of comfort or security obtained in current lifestyles, with the exception of individual mobility in the absence of major technological breakthroughs (see Girardet and Mendoça, 2009).

[16] See discussions in chapters III and V of this report.

[17] A further breakdown of this amount shows that $126 billion were spent on oil subsidies, $85 billion on natural gas, $6 billion on coal gas and $95 billion on fossil fuels for electricity generation.

[18] The amounts spent on these subsidies vary significantly from year to year, given the volatility in oil prices.

[19] Cogeneration of technologies refers to the possibility of developing new (but complementary) sets of technologies in parallel.

[20] An innovation system is defined as a network of economic and non-economic actors, the interactions amongst whom are critical for collaborative learning and application of knowledge to the creation of new products, processes, organizational forms, among others.

[21] See section 1(d) and particularly 1(d)(i) of the Bali Action Plan, available at: www.unfccc.int/resource/docs/2007/cop13/eng/06a01.pdf

[22] Or 1,163 kWh.

[23] See Energy Sector Management Assistance Programme (ESMAP, 2008) for an interesting study of approaches to maximize productive impacts of access to electrification projects.

[24] The importance of integrating poverty reduction in discussions on the green economy and RETs is becoming increasingly clear. For example, the UNEP defines the green economy as one ..."[t]hat results in improved human well-being and social equity, while significantly reducing environmental risks and ecological scarcities".

[25] "Clean technologies", or "clean energies", is generally a much broader concept than RETs, and includes clean coal.

REFERENCES

AGECC (2010). *Energy for a Sustainable Future: Summary Report and Recommendations*. New York, United Nations, April.

Balachandar G (2011). India closely follows China, US in wind power capacity addition. Available at: http://www.mydigitalfc.com/news/india-closely-follows-china-us-wind-power-capacity-addition-752 (accessed 8 August, 2011).

Birol F (2007). Energy economics: A place for energy poverty in the agenda? *The Energy Journal*, 28(3): 1-6.

Ernst & Young (2011). Renewable energy country attractiveness indices, no. 28, February, available at: www.energy-base.org/fileadmin/media/sefi/docs/publications/EY_RECAI_issue_28.pdf.

ESMAP (2007). Technical and economic assessment of off-grid, mini-grid and grid electrification technologies. Technical Paper no. 121/07, Washington, DC, World Bank Group, December.

ESMAP (2008). Maximizing the productive uses of electricity to increase the impact of rural electrification programs. Technical Paper no. 332/08, Washington, DC, World Bank Group, April.

Foster V and Briceno-Garmendia C (eds.) (2010). Africa's Infrastructure .A Time for Transformation. A co-publication of the Agence Française de Développement and the World Bank http://webcache.googleusercontent.com/search?hl=en&q=cache:rMN_kUGQeMQJ:http://siteresources.worldbank.org/INTAFRICA/Resources/aicd_overview_english_no-embargo.pdf+Foster+and+Briceno-Garmendia+2010&ct=clnk

Girardet H and Mendoça M (2009). *A Renewable World: Energy, Ecology, Equality.* A report for the World Future Council. Totnes, Devon, Green Books.

Holm D (2005). Renewable energy future for the developing world. ISES White Paper, International Solar Energy Society, Freiburg.

IEA (2009). *World Energy Outlook 2009*. Paris, OECD/IEA.

IEA (2010). *World Energy Outlook 2010*. Paris, OECD/IEA.

IPCC (2008). *Climate Change 2007. Synthesis Report*. Geneva, Switzerland.

IRENA (2011) forthcoming. Scenarios and Strategies for Africa. IRENA-Africa High-Level Consultations on Partnership on Accelerating Renewable Energy Uptake for Africa's Sustainable Development. Abu Dhabi.

Martínez DM and Ebenhack BW (2008). Understanding the role of energy consumption in human development through the use of saturation phenomena. *Energy Policy*, 36(4): 1430–1435.

Monari L (2011). Presentation Africa energy access challenges. IEA, Paris, 13 May 2011.

Nordhaus WD (1994). Do real output and real wage measures capture reality? The history of lighting suggests not. New Haven, CT, Cowles Foundation for Research in Economics, Yale University, September.

Payne JE (2010). Survey of the international evidence on the causal relationship between energy consumption and growth. *Journal of Economic Studies*, 37(1): 53–95.

REN21 (2010). *Renewables 2010*. Global Status Report. Paris, REN21 Secretariat, September.

Stern NH (2007). *The Economics of Climate Change: The Stern Review*. Cambridge, Cambridge University Press.

UNCTAD (2010a). *World Investment Report: Investing in a Low-carbon Economy*. United Nations publication, sales no. E.10.II.D.2. New York and Geneva, United Nations.

UNCTAD (2010b). *The Least Developed Countries Report 2010: Towards a New International Development Architecture for LDCs*. New York and Geneva, United Nations.

UN/DESA, UNEP and UNCTAD (2010). *The Transition to a Green Economy: Benefits, Challenges and Risks from a Sustainable Development Perspective.* Report of Panel of Experts to the Second Preparatory Committee Meeting of the United Nations Conference on Sustainable Development. New York, United Nations. Available at: http://www.uncsd2012.org/rio20/index.php?page=view&type=400&nr=12&menu=45

UN/DESA (2011). *World Economic and Social Survey 2011: The Great Green Technological Transformation*. New York, United Nations Department of Economic and Social Affairs.

World Bank (2009). *World Development Report 2010 : Development and Climate Change.* Washington, DC, World Bank.

World Bank (2010). *Africa's Infrastructure: A Time for Transformation*. Washington, DC, World Bank.

2

RENEWABLE ENERGY TECHNOLOGIES AND THEIR GROWING ROLE IN ENERGY SYSTEMS

CHAPTER II

RENEWABLE ENERGY TECHNOLOGIES AND THEIR GROWING ROLE IN ENERGY SYSTEMS

A. INTRODUCTION

The need to expand access to energy in order to drive global growth and job creation while simultaneously producing fewer GHG emissions is becoming increasingly recognized. Renewable energy technologies (RETs), which can be mixed with conventional energy sources, could provide countries with varied energy options within their national energy matrices to suit their specific needs and conditions. Given their enormous potential, there is growing interest in the current and future role of RETs in national energy supply systems worldwide. The nature of RETs and their current and possible future role are examined in this chapter, thereby establishing the basis for the discussions of policies relating to RETs in the subsequent chapters.

RETs are a diverse group of technologies, and although there are problems of intermittency associated with some of them (for example, in the provision of solar energy, where sun is available only for a limited number of hours per day), they are very versatile in that they can be deployed in various configurations. Therefore they offer the potential to contribute significantly to alleviating energy poverty in diverse situations. They can either be applied alone or, often, in combination with conventional energy technologies. They offer flexibility in their scale of application, from very small to very large, ranging from non-grid-based to semi-grid and large-scale grid applications. Because of their possibility of use in non-grid or semi-grid applications, RETs can be an important means of energy supply in areas where other energy sources are not available, such as in isolated rural communities. Such decentralized, off-grid applications of RETs are already in relatively wide use in developing countries, where they provide significant benefits to local communities (UNCTAD, 2010). While some of these applications are small in scale and do not make much of an impact on energy provision at the national/global level, they can still play an important role in reducing energy poverty at the local/rural level. The benefits of decentralized applications can be very large relative to absolute amounts of energy provided, because the marginal utility of the first few units of electric power (in particular) are much higher than the marginal utility of additional units of power for those who already have access to national grids. In other words, the value of gaining some access to energy and the social returns from that access for a severely energy-deprived population which currently has little or no access are likely to be very high. Also, RETs can be configured in many ways to provide energy on a larger scale thereby making a sizeable contribution both to meeting global energy needs and to mitigating climate change.

RETs are very versatile and can be deployed in various configurations.

RETs can be an important means of energy supply in areas where other energy sources are not available, because of their possibility of use in non-grid or semi-grid applications.

Countries with abundant RE sources have considerable potential to tap into them for augmenting national energy supply. The most mature and widely deployed RETs are based on hydropower, biomass, wind and solar energy. They are also the fastest growing, while several other RETs are in their early stages of development. In most scenarios on the role of RE sources in global primary energy supply by 2030 and 2050, three RETs are expected to make the

Future expansion of RETs and their contribution to global energy supply will depend on further technological progress... and on national and international policy choices.

RETs are a diverse set of technologies that convert renewable energy sources into usable energy in the form of electricity, heat or fuel.

largest contribution: modern biomass, wind and solar (IEA, 2010a; IPCC, 2011). However, the extent of future expansion of RETs and their contribution to global energy supply will depend partly on further technological progress leading to greater cost reductions in their use. It will also largely depend on national and international policy choices in the coming years. These choices relate to measures that level the playing field and have to do with fossil fuel subsidies, incorporating externalities not currently captured by market prices for energy by establishing a price for carbon, promoting additional investments in RETs and improving energy infrastructure, policy support to RE technology transfer, diffusion and absorption among countries, and ensuring effective financing mechanisms to enable such deployment, especially to the poorer developing countries and LDCs (as discussed in chapters III–V).

Section B of this chapter starts with a discussion of the nature of RETs, their characteristics and the diverse configurations in which they can be applied, as well as their role today and in the future as alternative sources of energy. Section C presents trends in private and public investment in RETs globally, and discusses the key issue of the high costs of RETs compared with conventional sources of energy.

B. DEFINING ALTERNATIVE, CLEAN AND RENEWABLE ENERGIES

The term "alternative energy" is generally intended to mean alternatives to fossil fuels. In some reports the terms *renewable* and *clean* energy are used interchangeably. However, for the purposes of this report, RETs differ from clean energy technologies (CETs) and "alternative" energy technologies, as defined below.

CETs are usually defined as those energy-generating technologies that have the potential to reduce GHG emissions (UNEP, EPO and ICTSD, 2010). They emit relatively little carbon dioxide (CO_2) or other GHGs, even though they may rely on non-renewable inputs, require significant waste disposal and/or pose the risk of "dirty" accidents. A major example is nuclear power, which is relatively clean in terms of GHG emissions, but is based on the fission of uranium, which is a scarce natural resource. Nuclear waste is also highly toxic and difficult to store, and nuclear accidents can lead to the spread of health- and life-threatening radioactive materials. A narrower definition of "clean energy" than the current one might therefore exclude nuclear energy from the group of CETs. In terms of GHG emissions, natural gas is "cleaner" than coal and oil. "Clean coal", defined as manufactured gas or liquids, or even electric power, is based on a process that incorporates carbon capture and storage (CCS), and is thus much lower in net GHG emissions than "raw" coal. Therefore, it is also often considered to be a clean energy source. However, CCS technologies are intrinsically very energy-intensive, and have yet to be applied effectively on a large scale.

There is no universally accepted definition of renewable energy. Broadly speaking, it is energy derived from naturally replenishable sources (box 2.1).[1] For purposes of this Report, RETs are a diverse set of technologies that convert renewable energy sources into usable energy in the form of electricity, heat or fuel. The main renewable energy sources are flowing water (hydropower), biomass and biofuels, solar heat, wind, geothermal heat and ocean energy.

Most of the discussion on RETs in the literature, and in this Report, relates to electricity generation, either in central or decentralized facilities. Nevertheless, transport, industry, agriculture and housing account for a large part of global energy consumption, and there are non-electric RET applications in all of them, such as biofuels for transportation, space heating, hot water and cooking (e.g. by solar cookers). While the world economy appears to be electrifying slowly but surely, it is important to bear in mind that electrification – and access to elec-

Box 2.1: Definition of renewable energy

Renewable energy has various definitions. It has been defined as energy obtained from the continuous or repetitive currents of energy recurring in the natural environment, or as energy flows that are replenished at the same rate as they are "used" (Sorensen, 2000).[a] The IPCC defines RE as any form of energy from solar, geophysical or biological sources that is replenished by natural processes at a rate that equals or exceeds its rate of use (IPCC, 2011: 10). The rate of replenishment of these sources needs to be sufficiently high for them to be considered renewable sources by energy and climate policies. Therefore, fossil fuels (e.g. coal, oil, natural gas) do not fall under this definition. As long as the rate of extraction of the RE resource does not exceed the natural energy flow rate, the resource can be utilized for the indefinite future, and may therefore be considered "inexhaustible." However, not all energy classified as "renewable" is necessarily inexhaustible (Boyle, 2004).

Source: UNCTAD.

[a] This definition has been in use since the 1980s (see, for example, Twidell and Weir, 1986).

tricity – may occur in a decentralized form which does not require the universal availability of "central" power from large plants delivered via a "grid".

Furthermore, there is no standard classification of RE sources and technologies. The IPCC (2011) categorizes them as bio-energy, direct solar, geothermal, hydropower, ocean and wind. Bio-energy includes biomass and biofuels. However, some analysts exclude biofuels, while others categorize biomass and biofuels separately. UNCTAD (2010) adopts the classification used by the International Energy Agency (IEA, 2007, annex 1), which subdivides ocean energy into waves, tides and "other", and includes a separate category for combustible wastes, as well as the standard set (wind, solar, biomass and geothermal) already mentioned.

This Report does not discuss large hydropower and biofuels in detail. Large hydropower is a very mature technology with limited, short-term growth potential, *except* in remote locations, but it often requires the displacement and relocation of large numbers of people at great social and economic cost. In many cases, those people are self-sufficient tribal or rural societies that are moved away from their ancestral lands. Large hydroelectric projects can also have serious impacts on the ecosystem. Similarly, in the case of biofuels, linkages may not always be positive and may compete with other needs. These will need to be balanced in national contexts, taking into consideration the different aspects involved.

The focus of this Report is primarily on RETs based on wind, solar and modern biomass sources. These are among the most important and fastest growing RETs in developing countries (figure 2.1 shows the status in 2010). Much of the energy from solar photovoltaic (PV) installations in developing countries is generated off-grid, thus the data in figure 2.1 may be an underestimation of its actual use in those countries. Biofuels are used mostly as alternative fuels for automobiles, trucks and buses. In addition, solar, wind, wood (as chips or sawdust), agricultural waste (e.g. bagasse) and biogas can also supply primary energy for decentralized as well as centralized electric power generation.[2]

Electrification – and access to electricity – may occur in a decentralized form which does not require the universal availability of "central" power.

1. The growing role of RETs in energy systems

The supply of energy by RETs has risen rapidly over the past decade, especially since 2003 when hydrocarbon prices began surging. However, RETs (excluding large hydro-based technologies) still account for a relatively small fraction of global energy capacity and supply because they started from a very small base of installed capacity. This section discusses the current role of RETs globally and how that role may expand in coming decades. This is followed by the cost issue, which will strongly influence the speed and extent of their deployment globally.

In 2008, RE sources (including large hydro installations) accounted for 12.9 per cent of

The focus of this Report is primarily on RETs based on wind, solar and modern biomass sources…the fastest growing RETs in developing countries.

Figure 2.1: Renewable electric power capacity (excluding hydro), end 2010

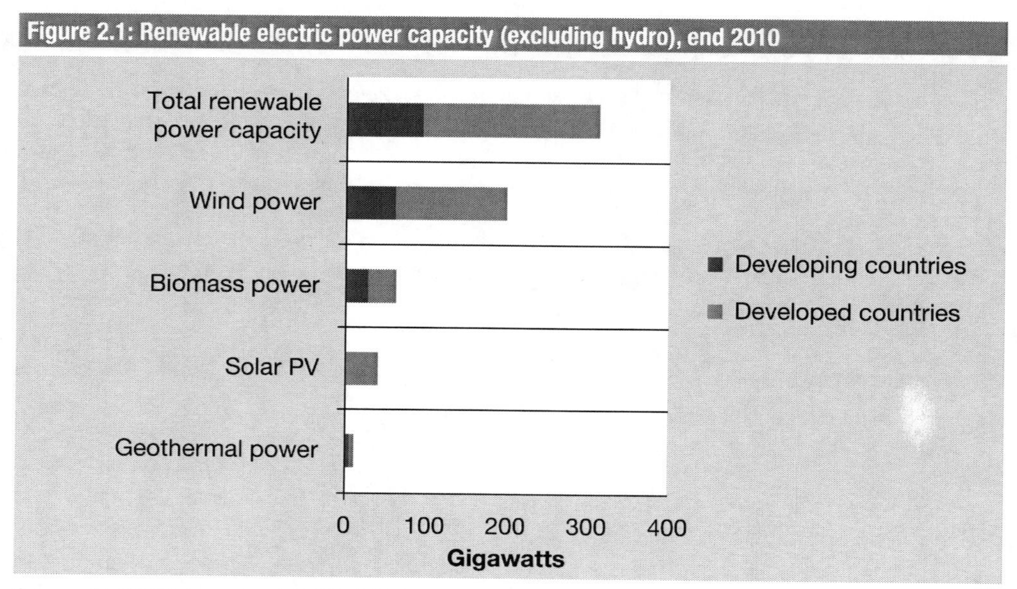

Source: UNCTAD, based on REN21 (2011).

Note: Estimates of electric power generation by solar PV installations in developing countries are from REN21 (2010). Other technologies not included in the chart, such as solar thermal power and ocean (tidal) power, present low levels of generation capacity: 1.1 and 0.3 GW respectively.

global primary energy supply (IPCC, 2011), whereas the bulk was supplied by fossil fuels (including oil, gas and coal). An estimated 21,325 TWh of electricity was generated in 2010 (REN21, 2011), of which 19.4 per cent was contributed by RE (figure 2.2), mainly in the form of hydropower (16.1 per cent) and primarily from large hydro installations. Nuclear power accounted for 13 per cent of the total in 2008. The share of fossil fuels in electric power generation increased slightly, not only accounting for the largest share of global energy capacity, but also constituting the main source of electricity in 2010 at 67.6 per cent of the total (REN21, 2011).

On a global scale, therefore, modern RETs today still supply only a small proportion of overall energy demand, despite very rapid growth of deployment in recent years. However, the total potential RE resources available globally are greater than total global energy demand, implying that there is much more potential to harness RE in the short to medium term through full implementa-

In 2008, RE sources (including large hydro installations) accounted for 12.9 per cent of global primary energy supply.

Figure 2.2: Global electricity supply by energy source, 2010

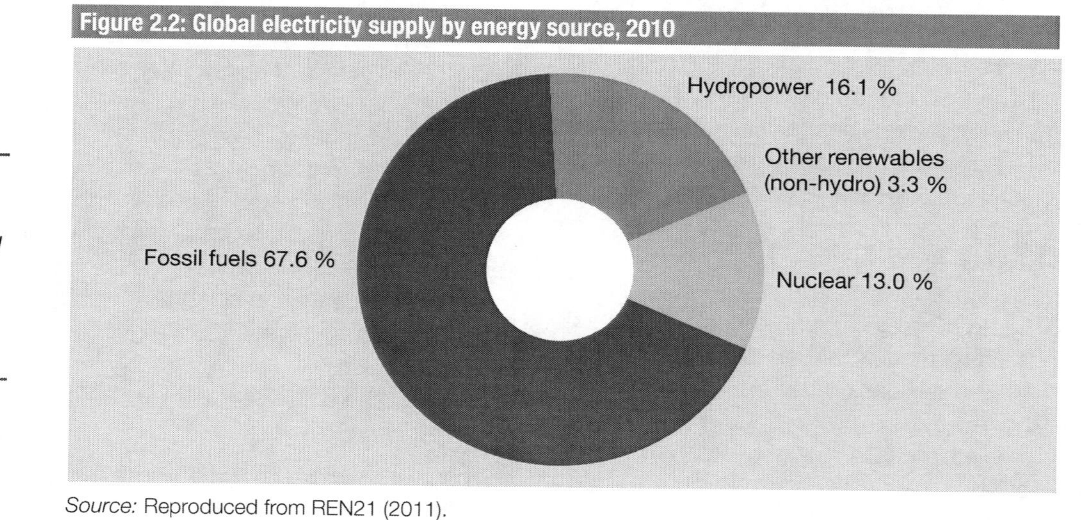

Source: Reproduced from REN21 (2011).

tion of demonstrated technologies or practices (i.e. without any new technology than is currently being utilized).[3] The availability of RE sources may differ greatly from the technical potential of such resources, which is the amount of renewable energy output that is theoretically obtainable through the full implementation of demonstrated technologies, regardless of costs, legal or other barriers and policy issues. The estimated technical potential of geothermal or wind power alone exceeded the global demand of 2008. The technical potential of RE for heating is also huge (IPCC, 2011). Against this background, the question arises as to how much of this technical potential will actually be harnessed in the future. How fast is power generation from RE and from the deployment of RETs growing? And what are the barriers to their wider deployment? In this context, the following five trends are worthy of note.

First, the relatively modest current contribution of RETs to global energy supply obscures the fact that some RETs have been growing very rapidly in recent years. During the period 2005–2010, for example, grid-connected solar PV technologies grew the fastest (at an average annual rate of 60 per cent), followed by all solar PV (49 per cent) and biodiesel production (38 per cent). Growth in the solar PV market accelerated still further in 2010, with the rapid decline in PV module prices in 2009, which made this technology more affordable and stimulated additional demand, particularly for small-scale, distributed generation projects, such as roof-mounted PV systems (REN21, 2010; World Economic Forum, 2011). There was also rapid growth in wind power (27 per cent), followed by concentrating solar power (CSP, by 25 per cent), ethanol production (23 per cent) and solar hot water/heating (16 per cent). By contrast, hydropower and geothermal power grew at modest rates (3–4 per cent) over the same period (REN21, 2011). Taking longer time periods, from 1971 to 2000 wind power grew 52.1 per cent, while solar grew by 32.6 per cent (Aitken, 2003). However, even with rapid deployment, it will take considerable time and investment in RETs for them to grow into major global sources of energy.

Second, in 2009, developing countries accounted for about half of all electric power generating capacity using RETs. The electricity generating capacity from RE (excluding large-scale hydropower) in developing countries has grown rapidly, almost doubling in five years, from 160 GW in 2004 to 305 GW in 2009 (REN21, 2005; and REN21, 2010). A detailed disaggregation by country or region is not possible due to data limitations; however, available data on global installed RE capacity in 2009 provide an indication of the breakdown by developed and developing countries. They show that developing countries accounted for over half (650 GW, or 53 per cent) of the total of 1,230 GW of RE electric power capacity (REN21, 2010: 55). China's share in the developing-country total was 35 per cent (or 246 GW), while India's was 4 per cent (or 49 GW). The 27 countries of the European Union (EU-27) accounted for 20 per cent of the global capacity (246 GW) and the United States for 11.7 per cent (144 GW). A major share of total RE capacity was from hydroelectric capacity from large-scale installations. At the end of 2010, excluding hydropower, 94 GW (or 30 per cent) of the renewable electric power capacity of 312 GW was located in developing countries (figure 2.1).

During the period 2005–2010, grid-connected solar PV technologies grew the fastest, followed by all solar PV and biodiesel production.

Third, in recent years, members of the Group of 20 (G-20) countries have accounted for most of the new investment in clean energy (part of which is RE) – reportedly 90 per cent of total investment in clean energy. China has invested particularly heavily in RE, and is also the fastest growing RE market. In 2010, it was the largest investor in clean energy, followed by Germany and the United States. Brazil and India have also been among the largest investors in clean energy in recent years. Over the period 2005–2010, the G-20 countries that have expanded clean energy investment the fastest in percentage terms included (in descending order) Turkey (with the highest),

In 2009, developing countries accounted for about half of all electric power generating capacity using RETs.

Argentina, South Africa, Indonesia, China, Brazil, Mexico and the Republic of Korea (Pew Charitable Trusts, 2011).

Fourth, RETs have already been deployed on a significant scale in some countries, though this varies by region. China, for instance, has the largest installed RE power capacity of all countries, and is further increasing that capacity. Over the period 2005–2010, RE capacity in China grew 106 per cent, followed by the Republic of Korea (88 per cent), Turkey (85 per cent), Germany (67 per cent) and Brazil (42 per cent) (Pew Charitable Trusts, 2011).

RETs have already been deployed on a significant scale in some countries, though this varies by region.

Finally, there is an increasing trend towards more deployment of RETs across the different regions. The following data on installed power capacity for wind, hydropower and geothermal power are indicative of recent deployment trends in RETs. Wind power capacity at the end of 2010 was the largest in Europe (86,075 megawatts (MW)), followed by Asia (58,641 MW), North America (44,189 MW), the Pacific (2,397 MW), Latin America and the Caribbean (2,006 MW) and Africa and the Middle East (1,079 MW) (GWEC, 2011).[4] Among developing countries, China (42,287 MW) and India (13,065 MW) have been the clear leaders in harnessing wind power. Among the developed countries, the United States (40,180 MW) was only slightly ahead of Germany (27,214 MW) and Spain (20,676 MW). Other developing economies with significant installed wind capacity include Turkey (1,329 MW), Brazil (931 MW), Mexico (519 MW), Taiwan Province of China (519 MW), the Republic of Korea (379 MW), Morocco (286 MW), Chile (172 MW), Costa Rica (123 MW) and Tunisia (114 MW) (GWEC, 2011). At least 49 countries added wind power capacity during the course of 2009 (REN21, 2010). In Africa, there has been less deployment, with total installed hydro RE capacity of 23 GW in 2009 (IPCC, 2011). There remains a large untapped potential, judging by the difference between the technical potential (i.e. potential for installed capacity) for annual power generation and actual generation, or installed capacity. Africa has a particularly large untapped potential (with 92 per cent of the potential undeveloped) followed by Asia (80 per cent) and Latin America (74 per cent) (IPCC, 2011, table 5.1). Hydropower deployment has been extensive in both Asia and Latin America, where installed capacity was substantial by 2009 (402 GW and 156 GW respectively). Significant increases in hydropower capacity are in the project pipeline for 2011, much of it concentrated in developing and emerging economies (including Brazil, China, India, Malaysia, the Russian Federation, Turkey and Viet Nam) (REN21, 2010).

At least 49 countries added wind power capacity during the course of 2009.

Geothermal deployment has been significant in developing countries in Asia – and is expected to increase – but much less so in Africa and Latin America, where it is not projected to increase much by 2015 (IPCC, 2011: table TS4.1). Nearly 88 per cent of the total known geothermal capacity is located in seven countries: the United States (3,150 MW), the Philippines (2,030 MW), Indonesia (1,200 MW), Mexico (960 MW), Italy (840 MW), New Zealand (630 MW) and Iceland (580 MW) (REN21, 2010). However, some 70 countries reportedly had geothermal projects under development as of May 2010, and projects are being planned or are under way in East Africa's Rift Valley, including in Kenya, Eritrea, Ethiopia, Uganda and the United Republic of Tanzania (REN21, 2010). Despite this, there remains huge untapped potential to further expand their use in all regions and in all country groups.

2. Limits of RET applicability

RETs vary in terms of technical efficiency, the different scales of application (from micro to macro), the potential for combining different technologies, the potential for off-grid use, the level of maturity, the type of energy product (electricity, heat or fuel) and the cost of the useful energy that they produce. The level of maturity is important as it has relevance for whether applications can be customized or whether large-scale deployment of pre-manufactured units is possible. The most advanced RETs (wind and solar) suffer from two main drawbacks. One is the intermittent supply of energy due to natural cycles (e.g. solar power relying on sunlight, and wind power on wind), which presents challenges for their integration into energy systems (IPCC, 2011). The second is high initial capital costs, given the reluctance of banks to lend to "risky" projects, at any scale, as they are considered as depending on "unproven" technologies. This is especially problematic if coal or natural gas is readily available. Unfortunately, for most situations, fossil-fuel solutions still offer the lowest up-front costs. A diesel generator costs about $1,000/kilowatt (kW) of capacity, compared with $3,000/kW to $6,000/kW for low-head hydropower.[5]

The energy output from some RETs is variable, and to some degree unpredictable, over different time scales – from minutes to years (IPCC, 2011). Additional R&D investment could immensely improve energy storage technologies for wind and solar power, and for some other energy technologies, such as batteries for electric cars, or for use with "smart" electric power grids (box 2.2). It could also look into increasing the cost-effectiveness of RETs for greater use in developing countries. Currently, such R&D is ongoing, and some possible solutions to the intermittency problem have already been demonstrated in several different grid-connected applications. However, resolving the technological constraints of intermittency will become more important as wind and solar PV technologies increase as a share of total energy supplied through electric grids (REN21, 2010; Eyer and Corey (2010); Singer 2010). The experience of several OECD countries shows that intermittency becomes a major issue for integration of wind power into energy systems at around the point where RE accounts for 20 per cent of total average annual electrical energy demand (IPCC, 2011). Below this threshold intermittency is less of an issue: at low rates of wind (or solar) penetration, intermittency may be managed by relying on a mix of REs along with conventional sources. In general, the integration challenges associated with RE are contextual, site-specific and complex (IPCC, 2011). In situations where RETs are expected to supply a share greater than 20 per cent of the total energy generated, problems of intermittency will need to be resolved through the development of local storage capability and/or grid connections.

The successful integration of intermittent energy sources on a large scale in the future may require the development of "smart" electric grids that can better accommodate REs (box 2.2).

Additional R&D investment... could look into increasing the cost-effectiveness of RETs for greater use in developing countries.

Box 2.2: Developing "smart grids" to better integrate RE sources into energy systems

The electric power grid is a network of generating plants, cables, switches and transformers that form the transmission and distribution systems for electricity. The transmission system delivers electricity from power plants to substations, while the distribution system delivers electricity from substations to consumers. The grid can also include many smaller local networks. Both Europe and the United States are actively considering how to upgrade existing electric power grids into "smart grids".

In essence, a "smart grid" is a modernized electric grid with an improved ability to integrate intermittent energy sources, and to efficiently manage all the different types of energy sources that feed into the grid in order to efficiently meet variations in electricity demand throughout the day. It would facilitate the integration of small RE generators, such as solar PV home systems, as well as larger RE sources such as onshore and offshore wind farms and solar power plants. The network would be "smart" in the sense of delivering both reactive and interactive capabilities in transmission and distribution. It would integrate digital information technology into regional and local electricity distribution networks, thereby making the electric grid more reliable, resilient and secure. It would also enable better demand management and energy-efficiency gains by consumers and businesses, and, incidentally, facilitate the large-scale deployment of electric vehicles.

Sources: UNCTAD, based on Eyer and Corey (2010); Singer (2010); Pollin, Heintz and Garrett-Peltier (2009) and various press reports.

The ease with which RETs can be integrated into existing energy systems will affect the rate of future deployment.

In the absence of electricity storage, electric utilities are required to match output to demand at all times. Gas turbines, stored hydro and geothermal power can be activated quickly, and some coal-fired power plants may be kept running in "spinning reserve" to respond quickly to surges in demand or to supply interruptions elsewhere in the system. However, nuclear power is very inflexible and operates best at full capacity. Wind farms, large solar PV farms, wave and tidal stations provide only intermittent power, generating electricity only when conditions are favorable. Isolated rooftop PV installations can be combined with batteries, or they can be designed to feed energy back to the grid (encouraged by so-called "feed-in" tariffs, as discussed in chapter V).

The ease with which RETs can be integrated into existing energy systems will affect the rate of future deployment. Many different energy systems exist globally, each with distinct technical, market and financial differences. Integration issues can be system-specific and resource-related (IPCC, 2011), such as rapidly dispatchable RE-based resources (especially gas turbines or stored hydropower). These may offer extra flexibility for the system in terms of its ability to integrate different RE sources (wind or solar PV in particular). The issue of intermittent energy supply is important because it affects the efficiency of generating power from existing installed capacity. Intermittent power supply is inappropriate for base-load requirements, and poses technical challenges to grid management. The difficulty of integrating intermittent renewable energies into electric grids can be reduced, to some extent, by improved real-time forecasting (on a time scale of minutes and hours) of likely variations in wind, for example, or of fluctuations in electricity demand. However, such integration necessitates large investments in energy infrastructure (IPCC, 2011). In any case, large investments would be needed to maintain and expand existing energy infrastructure in many countries, even in the absence of a scaling up of renewable energy resources.

Considerations of energy efficiency will play an important role in determining the possible extent of integration.

3. Established and emerging RETs

This section provides detailed descriptions of established and emerging RETs, including the characteristics and state of application of each RET, and does not limit itself to the three RETs that are the main focus of this Report, namely wind, solar and biomass. Considerations of energy efficiency will play an important role in determining the possible extent of integration of any particular RET into national energy mixes. Box 2.3 provides a simple explanation of energy efficiency issues.

a. Hydropower technologies

Hydropower technologies use power generated by harnessing the flow of water through a hydraulic turbine or equivalent. They vary greatly in the scale of generation capacity.[6] Small and large hydropower systems are the most mature of the RETs,

Box 2.3: Energy efficiency and conventional measures of thermodynamic efficiency

According to a convention that is now widely adopted by government and international energy agencies, *primary energy* is defined as the energy that is embodied in natural resources consumed by an economy (IPCC, 2011). Primary energy is transformed into *secondary energy* through cleaning (for natural gas), refining into petroleum products (for crude oil), coking (for coal) or by conversion into electricity, transport fuel or (useful) heat. Secondary energy that is delivered to an end-user, such as electricity supplied from an electrical outlet of a building, is called final energy (IPCC, 2011).

Each energy conversion involves some loss, characterized as *rejected energy*. For example, when primary energy (in the form of fuel) is converted to electric power, about two thirds of the primary energy is lost – or rejected – as low temperature heat.

Efficiency measures can be defined for each stage of energy transformation or conversion.

Source: UNCTAD.

and have been a relatively important source of electric power production for many decades in many countries (REN21, 2005). Large hydro accounts for the bulk of hydropower energy capacity. The technologies in installations of different sizes are not fundamentally different.

Theoretically, the total hydropower available globally has been estimated at 40,000 TWh per annum (WEC, 2010).Estimates of its technical potential for power generation range between 14,000 and 16,000 TWh (Boyle, 2004; WEC, 2010).[7] In 2008, hydropower accounted for about 16 per cent of global electricity supply and for 2.3 per cent of global primary energy supply (IPCC, 2011), and it was by far the largest RE contributor to electricity generation, although biomass contributes more to global primary energy supply.

Concerning the technical aspects, it is estimated that only 25 per cent of global hydropower potential has been developed. Most regions of the world have large untapped hydro resources, especially Africa with 92 per cent of its hydro resources undeveloped, but also South America and Asia (IPCC, 2011). Thus there is certainly scope for further development in these regions. Some developing countries have begun to invest into hydropower. Ethiopia, for instance, has formulated a 25-year national energy plan in 2005 to increase generation capacity from hydro resources, with the expectation that this will result in benefits for the economy in the medium to long term. The Government plan has so far resulted in an increase of 39 percent in generation capacity in the last five years: from 2,587 MW (2005) to 3981 MW (2010), most of which is attributable to hydropower. Although hydropower is a proven and well-advanced technology, some aspects of it could be improved further. Storage of hydro resources could be used to buffer mismatches between supply and demand, which is a valuable attribute.

As mentioned earlier in this chapter, large hydroelectric schemes are controversial for environmental and social reasons, because they rely on dams that can have negative social and environmental impacts. The construction of hydroelectric dams can also cause political and social conflicts between countries that share rivers and waterways because upstream dams may reduce water flow to downstream countries, either due to diversion into irrigation projects, excessive evaporation (e.g. the Aswan Dam in Egypt) or via seepage into the ground. However, despite these problems, evidence suggests that relatively high levels of deployment are feasible over the next 20 years (IPCC, 2011).

Small and large hydropower systems are the most mature of the RETs.

b. Biomass energy technologies

Biomass is biological material from either living or recently deceased organisms. It includes many types of plants and trees, as well as wood and waste, but is generally understood to exclude fossil fuels. Biomass energy technologies use both traditional and more sophisticated methods (referred to as *modern biomass power*) to produce useful energy primarily from wood residues, agricultural waste, animal waste and municipal solid waste. Such energy is derived from a variety of sources, including garbage and food scraps (yielding biogas), wood, municipal waste, landfill gases and alcohol fuels. Traditional biomass (wood and charcoal), "modern biomass" (i.e. collecting, pre-processing and delivering combustible cellulosic materials to electric power plants or chemical plants) and biofuels are three categories of biomass that are discussed below.

Biomass energy technologies use both traditional and more sophisticated methods (referred to as modern biomass power) to produce useful energy.

(i) Traditional biomass

Traditional sources of biomass, such as dead trees, tree branches and animal dung, have long been used in many developing countries for cooking and heating. The major energy conversion technology in rural communities consists of inefficient charcoal production followed by combustion of the char (or wood or dung) in simple cast iron or brick stoves or furnaces. While charcoal is often commercialized, traditional biomass, such as straw, tree branches or dung, is gathered without payment, largely by poor households for their own use as cooking fuel. It is estimated that 2.7 billion people, mostly in Africa and Asia, still cook using traditional biomass, which explains why it is still the largest RE source of total global primary energy supply today. However, traditional biomass is considered to be an extremely inefficient energy source, because the charcoal is produced in primitive, open air kilns that consume most of the energy content of the fuel to drive off the moisture and volatile materials.

Traditional indoor uses of biomass (such as in crude stoves) are associated with various health problems caused mainly by indoor air pollution (as mentioned in chapter I). Improved cooking stoves, which increase energy efficiency and reduce indoor air pollution, are being used increasingly in developing countries, but their wider deployment is needed (see REN21, 2011). There are also social and gender issues involved, because young girls and women are often assigned the task of collecting biomass, and they may spend several hours a day walking long distances in search of it. This reduces the time available for their education, leisure and other activities. It can also lead to environmental degradation, because young plants are often harvested for fuel before they have a chance to grow (UNCTAD, 2010). Sustainability of traditional biomass supply is therefore an important concern for many developing countries. Energy production from traditional biomass may fall in the future, as more people gain access to other sources of energy that are less harmful and easier to harness.

2.7 billion people, mostly in Africa and Asia, still cook using traditional biomass, which explains why it is still the largest RE source today.

(ii) Modern biomass for electric power

Biomass can also be converted into energy through alternative methods that are more efficient and do not give rise to the health hazards and problematic social issues associated with traditional biomass. Agricultural, animal and human waste, as well as other organic waste, all release methane (also called biogas or landfill gas (LFG)) when they decompose.[8] The process works on any scale, but the larger the scale the more efficient it is likely to be.

Energy production from traditional biomass may fall in the future, as more people gain access to other sources of energy.

Biogas of a more sophisticated type can also be produced from cellulosic materials, such as agricultural waste, by a process called steam reforming. Commercial biomass energy technologies to produce electric power are now fairly widely available. Biomass power plants include biomass gasifier power systems, biomass steam electric power systems, and municipal waste and biogas electric power systems. There is also the possibility of cogeneration – for combined heat and power (CHP) production – whereby heat is harnessed (for heating purposes) at the same time as electricity is generated. Cogeneration plants therefore improve energy efficiency by making use of heat that might otherwise go waste. These plants can use biomass, geothermal or solar thermal resources (REN21, 2011), and are similar to conventional power plants that run on fossil fuels. Biogas power plants generally range in size from a few hundred kilowatts to as much as 100 MW, and they may even be larger in big cities.

The production of biogas depends on the supply of biomass, and, in principle, can therefore be controlled. In this respect it is similar to biofuels, but is different from most other REs, which generally depend more directly on natural energy flows to generate power. Intermittency is therefore less of an issue with biomass than with some other REs.

Modern forms of biomass such as wood chips or pellets are also being used increasingly in advanced heating applications such as home heating, especially in the countries of the European Union (EU) (REN21, 2011) and some other developed countries.

(iii) First and second generation biofuels

Biofuels are liquid fuels made from plant material that can be used as a substitute for, or as an additive to, petroleum-derived fuels. There are two types of biofuels: alcohols (ethanol, methanol or butanol) and biodiesel. Ethanol, which is by far the most commonly used of the alcohols, is typically added to gasoline in a ratio of about one part to ten. Biodiesel, which is an oil made from oilseeds, can be mixed with conventional, petroleum-based diesel oil, or in some cases it can replace petroleum-based diesel fuel altogether.

Biofuels may be classified as either "first generation" or "second generation". First-generation biofuels – those that are currently available commercially – are produced from edible food grains, seeds and sugar crops. Ethanol and butanol are produced from sugar cane (Brazil), sugar beets (Europe) or corn (United States). Ethanol from sugar cane is probably the most attractive option among the first-generation biofuels since it is cheaper to produce (UNCTAD, 2008).

Biofuels compete with agricultural produce in two important ways. First, they may be based on edible biomass, which means that many of the primary sources can be used either as food (or feedstock) or as fuel, resulting in direct competition between the two uses. This is generally the case with first-generation biofuels, which are based on edible biomass. This competing use has led to controversy over their potential to reduce the availability of food and to raise food prices, thereby contributing to food insecurity and food crises (see, for example, Ford Runge and Senauer, 2007). Concerns have also been raised about the relatively low net energy output from many first-generation biofuels, as well as environmental impacts resulting from the large-scale use of water and fertilizers to produce them. In some cases, the GHG abatement levels from biofuels have also been criticized as being low.

Even when biofuels are not based on edible biomass (as in the case of second- generation biofuels), there could still be competition between producing material for biofuels and for food production in terms of land use (in quantity and quality) and water use.

A potentially promising new approach is the use of algae, grown either in fresh water ponds or in salt water. In contrast to agricultural crops, algal ponds can, in principle, produce as much as 40,000 to 80,000 liters of biodiesel per acre (Briggs, 2004). However, so far it has been difficult to control the algae growth process adequately to produce a continuous output of feedstock for a refinery, and it is not clear whether these difficulties are fundamental or temporary. In fact, a number of algae production companies have failed.

It is evident that ethanol, methanol and biodiesel do not currently qualify as negative cost (profitable) solutions. However the long-term potential for second-generation technology is quite favourable. The prospects for second-generation biofuels will be clearer in two or three years time, as a result of recent biotech breakthroughs that are expected to increase the productivity of conversion processes, which are important for extracting biofuels from biomass.

Even when biofuels are not based on edible biomass, there could still be competition between producing material for biofuels and for food in terms of land and water use.

c. Wind energy technologies

Wind energy technologies, mainly wind turbines, use kinetic energy from air currents arising from uneven heating of the earth's surface to generate electricity. The wind turbines, which are usually operated in groups in the form of a wind farm, wind project or wind power plant, are interconnected to a common utility system through a system of transformers, distribution lines and (usually) one substation. There are also hybrids such as small wind turbines combined with diesel generators or with solar PV panels.

Wind energy technologies are more standardized than solar technologies. The variations are mainly in terms of the size and location of the units. The two main classes are *onshore* and *offshore*. To date, offshore wind turbine designs have been very similar to onshore designs, but they tend to be larger and need special foundations to resist the wave action. Wind turbines can be applied off-grid or on-grid, though the larger projects are generally grid-connected. It is the larger, grid-connected wind farms, mainly onshore, that provide the bulk of electricity generation from wind, although there is considerable potential for future development offshore as well.

Wind turbines can be applied off-grid or on-grid, though the larger projects are generally grid-connected.

Wind turbines are rapidly being deployed, but face several challenges. Wind power

Intermittency is the major problem with wind energy...but the integration of wind energy into energy systems poses no insurmountable technical barriers.

is location-specific, which constitutes a significant limitation on its application. Intermittency is the major problem with wind energy, as the wind does not blow continuously and the electrical output of wind power plants varies with fluctuating wind speeds. The predictability of wind speeds is also an issue, as fluctuations are very difficult to predict, even a few seconds in advance, especially in the case of wind gusts. Storage remains a notable problem for wind energy (as in the case of solar PV). The state of storage technology is discussed in a separate section, later in this *TIR*. Still, experience and detailed studies from many countries have shown that the integration of wind energy into energy systems poses no insurmountable technical barriers (IPCC, 2011).

Onshore wind energy systems are relatively standard, whereas offshore wind energy technology is less well developed, and investment costs are generally higher. Onshore wind energy can be competitive with conventional energy sources, while offshore wind energy is currently relatively expensive. This is due to the comparatively less mature state of the latter technology, and because of the greater logistical challenges of maintaining and servicing offshore turbines (IPCC, 2011). (In this respect, they share similar maintenance challenges with turbines used in some ocean energy technologies.) Wind power can be distributed over existing networks when they are available nearby. However, for offshore wind and for remote onshore locations, existing transmission networks usually need to be extended. The distance from large population centers and energy consumers determines the amount of extension needed in the transmission network and the associated cost, and therefore varies from case to case.

Over the past three decades, innovation in wind turbine design has led to significant cost reductions.

Onshore wind energy technologies are mature, having been in use for several years. However, the use of wind energy to generate electricity on a commercial scale became viable only in the 1970s, starting in Denmark, as a result of technical advances and government support (IPCC, 2011). It is now being widely deployed internationally. Indeed, wind energy is now established as part of the mainstream electricity industry in many developed countries. However, existing wind power capacity remains regionally concentrated, with most existing capacity in Europe, North America and East Asia (REN21, 2010). China has been actively scaling up its wind power capacity. At least 82 countries use some wind energy on a commercial basis, but countries in Latin America, Africa, West Asia and the Pacific regions have installed relatively little wind power capacity to date, despite its significant technical potential.

Over the past three decades, innovation in wind turbine design has led to significant cost reductions. Newer designs utilize lighter materials (such as those used in aircraft) and compact generators with powerful permanent magnets based on the iron-boron-neodymium alloy.[9] Modern, commercial grid-connected wind turbines have evolved from small and simple to larger, highly sophisticated devices. Scientific and engineering expertise and advances, as well as improved computational tools, design standards, manufacturing methods and operating and maintenance procedures, have all supported technological progress and learning. In order to reduce the levelized cost of electricity (LCOE) from wind energy, typical wind turbine sizes have grown significantly. The LCOE of a particular project is defined as the constant price per kWh that electricity would have to be sold at in order for the project to break even over its lifetime. Many onshore wind turbines installed in 2009 have a rated capacity of 1.5 MW to 2.5 MW, while that of offshore installations goes up to 5 MW. Larger scale manufacturing of wind turbines is expected to reduce their cost still further.

Although wind resources depend on geography and are not evenly distributed worldwide, in most regions of the world the technical potential exists to enable significant wind energy deployment. According to the Global Wind Energy Council (GWEC), at the end of 2010 there was 194,390 MW of wind power capacity installed globally. Most of the installed capacity was in Europe, Asia and North America. The bulk of offshore wind capacity is located in Europe (REN21, 2010). The wind power capacity installed by the end of 2009 was capable of meeting roughly 1.8 per cent of worldwide electricity demand. According to the IPCC (2011), that contribution could grow to over 20 per cent by 2050 under some scenarios. A growing number of global wind resource assessments have demonstrated that global technical potential exceeds current global electricity production (IPCC, 2011).

d. Solar energy technologies

Solar energy technologies capture energy from the sun either as heat or as electricity through conversion by solar PV panels. There are three main classes of solar energy technologies: concentrating solar power (CSP) systems; solar thermal systems for heating residential and commercial buildings (which can be either active or passive in nature) and solar PV power systems.

(i) Concentrating solar power systems

Concentrating solar power systems are highly sophisticated and land-intensive, and are suitable mainly for desert areas where the sun shines during most daylight hours. The idea is to use either lenses or mirrors on the ground to capture solar energy that can be focused on a small receiving unit.[10] Moreover, because the period of maximum heat capture (midday) does not necessarily correspond to peak heat or electricity demand, either the heat captured or the electrical power produced needs to be stored.

An important advantage of CSP technologies (except for small units using parabolic dishes with Stirling engines) is their ability to store thermal energy after it has been collected at the receiver and before it goes to the heat engine. The majority of CSP plants in operation today rely on parabolic trough technology. This is however expected to change. Nearly half of the capacity in construction or under contract uses or will use linear Fresnel, dish/engine, or power-tower technology in the near future (REN21, 2010).

Costs of power from existing systems range from 19 cents/kWh to 29 cents/kWh. With increased plant sizes, better component production capacities, more suppliers and improvements through R&D, costs could fall to a range of 15 cents/kWh–20 cents/kWh (Greenpeace International, SolarPACES and ESTELA, 2009). Optimists believe that costs of CSP could decline rapidly to equal the cost of power from gas-fired power plants in 5 to 10 years. Under the most optimistic scenario, CSP could provide 18–25 per cent of global electricity needs by 2050, depending on the degree of improvements in energy efficiency achieved (Greenpeace International, SolarPACES and ESTELA, 2009).

An important advantage of CSP technologies is their ability to store thermal energy after it has been collected.

Spain and the United States are currently the leaders in CSP installed capacity, although several developing countries (including Abu Dhabi, Algeria, China, Egypt, Jordan, Morocco, South Africa and Tunisia) have begun to use CSP or have announced plans for CSP projects (IPCC, 2011). In North Africa, the Desertec project, supported largely by German firms, envisages a total investment of €400 million. It could not only provide a considerable share of Europe's demand for electricity, but could also constitute a major new industry for the Maghreb region of North Africa, especially Algeria and Morocco.[11]

Costs of CSP could decline rapidly to equal the cost of power from gas-fired power plants in 5 to 10 years.

(ii) Solar thermal systems

Solar thermal heating for buildings is a relatively old technology. However, it is increasingly important for new buildings, especially the "passive house" designs, which rely entirely on solar heat, combined with insulation, double- (or triple) glazed windows, and counter-current heat exchangers to heat incoming ventilation air (mainly to kitchens) from the outgoing air. In the summer, this procedure can be reversed to conserve air-conditioning (Elswijk and Kaan, 2008). In new construction, this technology can cut energy consumption by 90–95 per cent, although the capital costs are roughly 10 per cent higher than conventional designs. The major challenge is to find ways to retrofit some of these efficiency gains in existing buildings at reasonable costs.

Unfortunately, the benefits that can be achieved from fairly obvious leak reductions, such as insulating cavity walls, roofs and double-glazing single-glazed windows, are far fewer than can be achieved by new construction with air-exchange systems. According to one study, efficiency gains from insulation add-ons which are a fairly typical example – would probably be about 25 per cent (Mackay, 2008). However, lower thermostat settings can probably achieve about the same level of improvement at no (or negative) cost.

Solar thermal heating for buildings... can cut energy consumption by 90–95 per cent.

(iii) Solar photovoltaic technology

Solar PV is a semiconductor technology that converts the energy of sunlight (photons) directly into electricity. This technology has been known for a long time, but the first commercial applications utilized ultra-pure scrap silicon from computer chip manufacturing. However, the purity requirements for silicon PV cells are much less stringent than for chips, and by 2000 the demand for solar PV systems justified investment in specialized dedicated fabrication facilities.[12]

Solar PV power systems can be either off-grid or connected into mini-grids or larger national grids. Grid-connected systems may be either distributed or centralized. The distributed version consists of a large number of small local power plants, some of which supply the electricity mainly to on-site customers (such as houses), while the surplus (if any) feeds the grid. The centralized system works as one large power plant. Off-grid systems are typically dedicated to a single or small group of customers, and generally require an electrical storage element or back-up power (IPCC, 2011). In addition to the various solar PV technologies, there are various hybrids, including solar-PV wind hybrids, and solar and conventional mixed hybrids.

Solar PV power systems can be either off-grid or connected into mini-grids or larger national grids.

Intermittency is a major problem with solar PV, which operates only when the sun is shining. As with wind, storage for off-grid applications is possible with storage batteries, but the economics is unattractive. Thus intermittency remains a problem that has not yet been resolved and requires further technological progress. In contrast, storage is possible with CSP, although CSP with thermal storage is more costly than CSP without it.

Even for grid-connected solar PV systems, local output varies, not only predictably according to the diurnal (night and day) cycle, but unpredictably according to weather conditions. In some instances, this variability can have a significant impact on the management and control of local transmission and distribution systems, and may constrain integration into the power system. Predictability also varies with location, although in many cases where there is a high level of solar exposure there is also a reasonably high level of weather predictability. In any case, solar PV applications are expanding fast and the technology is still developing rapidly, resulting in lower prices (box 2.4).

Investments in production capacity in China may have significantly exceeded global demand, leading to a considerable oversupply, so that price-cutting has become endemic. Indeed, module prices fell by 25 per cent during the first half of 2011 (Photon International, 2011). Of the 400 or more solar PV firms in the world in early 2011, quite a large number are likely to fail in the coming months and years as the industry consolidates (Photon International, 2011).

Solar PV technology is currently being deployed at a rapid rate, including in some developing countries. In 2009, installations for RE generation worldwide were 7,000 MW (7 GW), of which half was in Germany. Total worldwide generating capacity at the end of 2009 was 22 GW, of which 9 GW was in Germany, followed by Spain, Japan and the United States (IPCC, 2011).

Box 2.4: Prices, production and capacity of PV systems

The prices of PV modules have been declining worldwide as a result of progress in manufacturing (mainly larger scale production), strong investment interest accompanied by policy support, tariff digressions and particularly technological progress in cell architecture. These changes have triggered the spectacular rise in global production of PV modules over the past decade. In 2010, a total estimated of 17 GW capacity has been added to the system. This represents a significant rise in total new capacity, when compared to the 2009, where 7.3 GW capacity was added globally. On the whole, the total global capacity in 2010 (approximately 40 GW) represents a 6 fold increase over what was observed in 2006 (7 GW) (REN21, 2010).

Most of this increase in capacity is attributable to Spain, Japan, United States and China, which already stands out as one of the largest producers. In terms of share of the global solar PV capacity, five countries represented about 80 per cent of the total available capacity, i.e. Germany (44 per cent), Spain (10 per cent), Japan (9 per cent), Italy (9 percent) and United States (6 per cent) (REN21, 2011: 23).

Source: UNCTAD, based on REN21 (2010) and REN21 (2011).

Solar PV is a versatile technology for reducing severe energy poverty and providing some degree of access to energy, even in remote villages. While it is true that the power is intermittent and available mostly during the daylight hours, battery storage for a few hours or days is feasible, and some important uses (such as refrigeration) can also be effective with intermittent power.

Solar power is being used in many innovative ways on a micro scale to power applications such as cookers, water pumps and crop dryers, lights, radios and televisions. These applications can help to considerably improve the lives of energy-poor, isolated rural communities in many developing countries, even though the quantity of energy supplied may be of little significance relative to global energy consumption. They provide an effective tool for reducing energy poverty when national power grids are severely deficient and where large segments of the population remain unconnected to grids.

e. Geothermal energy technology

There are two geothermal technologies. The oldest "conventional" technology extracts energy from existing reservoirs of steam or hot water in porous rocks beneath the earth's surface. Indeed, the technology for electricity generation from hydrothermal reservoirs is mature and reliable, and has been operating for about 100 years. Geothermal reservoirs are located conveniently near the earth's surface in many places, such as Iceland, where they already provide 25 per cent of the nation's electricity. Iceland is probably the most promising location in the world for such RETs at present, though other locations exist along the "ring of fire" where volcanoes are located, such as Indonesia, Japan, the Philippines, the west coast of South America, and parts of the western United States, as well as Greece, Italy and Turkey.

Solar PV is a versatile technology for reducing severe energy poverty and providing some degree of access to energy, even in remote villages.

Box 2.5: Geothermal energy: technical aspects

The main system, called engineered geothermal system (EGS), consists of a pair of pipes drilled into a bed of hot dry rock which is then fractured by water injection. (The fracturing process can cause small earthquakes, which will not be popular among local residents.) The resulting steam is then brought to the surface through the second pipe and used in steam turbines to generate electricity, or in heating equipment to provide industrial heating. At present, the steam from the exhaust is discarded into ponds, but with improved technology it could be reinjected, both to maintain internal pressure and to extend the life of the system. At present, although a number of projects are under development, the technologies for EGS are still at the demonstration stage. The main types of direct geothermal applications include space heating of buildings, bathing and balneology, horticulture (greenhouses and soil heating), industrial process heat, agriculture, aquaculture (fish farming) and snow melting.

Source: UNCTAD, based on IPCC (2011) and Boyle (2004).

Heat can also be reached by drilling deep enough at any location using an engineered geothermal system (EGS), and brought to the surface as hot water or steam to produce heat or electric power (box 2.5). In principle, hot dry rock from 3 to 10 kilometers (km) below the surface offers huge potential in the United States – about 140,000 times the energy consumption of the United States in 2007.

Geothermal energy is a stable, non-intermittent source and provides predictable energy supply throughout the day.

However, the operation of geothermal fields may provoke local hazards arising from natural phenomena, such as micro-earthquakes, hydrothermal steam eruptions and ground subsidence (IPCC, 2011). Geothermal plants also require a large up-front investment due to the need to drill deep wells and build power plants (of conventional design). Despite this, they incur low variable costs once the plants are constructed. Costs of power generation depend on the location of the geothermal reservoirs, which are sometimes far from large population centers and therefore require extension of the transmission network.

Geothermal energy is a stable, non-intermittent source and provides predictable energy supply throughout the day, unlike some other REs (Boyle, 2004). This makes geothermal technologies particularly suitable for base-load supply, but much less appropriate for isolated remote communities, except in very unusual circumstances. Integration of new power plants into existing power systems does not present a major challenge (IPCC, 2011).

The potential for additional energy supply from geothermal sources is reported to be high in a few areas where geothermal resources are plentiful. Recent statistics indicate that global geothermal power supply today amounts to 10,715 MW, enough to generate 67,250 GWh of energy. However, only 24 countries are currently using this source of energy (REN21, 2010; Johansson TB, 2011). In 2008, conventional geothermal energy use represented only about 0.1 per cent of the global primary energy supply. A number of start-ups in different countries are planning EGS projects, and, according to industry association forecasts, the total energy supply from geothermal sources will be 160 GW by 2050, about half produced by EGS. This implies that by 2050 geothermal could meet roughly 3 per cent of the global electricity demand and 5 per cent of the global demand for heating and cooling (IPCC, 2011).

f. Ocean energy technologies

Ocean energy can be defined as energy derived from technologies that utilize seawater as their motive power, or harness the water's chemical or heat potential. The RE from the ocean comes from five distinct sources: wave energy, tidal range (or tidal rise and fall), tidal and ocean currents, ocean thermal energy conversion (OTEC) and salinity gradient (or osmotic power). Each of them requires different technologies for energy conversion, and there is a considerable diversity of mechanisms involved. Moreover, since each type of ocean energy is driven from different natural energy flows, they each have different variability and predictability characteristics (box 2.6).

Box 2.6: Ocean energy technologies: technical aspects

Wave energy is generated by wind blowing over the sea. Wave energy technologies involve different physical structures (called wave energy converters) that move with the waves and convert their movement into usable energy. The movement back and forth of the tides can be harnessed in several ways to generate power. One way is through tidal mills, which are similar to watermills; another is by placing independent turbines in the tide; and a third is using a dam (called a tidal barrage) to trap the water and pass its flow through sluice gates that drive water through turbo-generators. These barrages can have environmental impacts similar to hydroelectric dams, discussed earlier. Also being developed are tidal current technologies that use the tidal currents to power underwater turbines, but they are very new and remain at an early stage of development (Boyle, 2004).

Source: UNCTAD, based on Boyle (2004).

Ocean energy technologies are by far the least mature of the major RETs (REN21, 2010). Most of them are still in their pre-commercial stages of development, ranging from the conceptual and pure R&D stages to the prototype and demonstration stage. Only tidal range technology using barrages can be considered mature, and is the only one commercially available so far. Currently there are several technology options for each ocean energy source, and, with the exception of tidal range barrages, technological convergence between these various sources has not yet occurred (IPCC, 2011).

These technologies appear to be relatively expensive, although the data on cost characteristics are not as well established as for other RETs. There are encouraging signs that the investment cost of ocean energy technologies and the LCOE generated from wave or tidal technologies will decline from their present levels as R&D and demonstrations proceed and as deployment occurs (IPCC, 2011). How far and fast these reductions take place will be a key determinant of their deployment in the future.

It is estimated that 6 MW of wave/tide energy systems is operational or being tested in Europe (off the coasts of Denmark, Italy, the Netherlands, Norway, Spain and the United Kingdom), with additional projects off the coasts of countries such as Canada, India, Japan, the Republic of Korea and the United States. At least 25 countries are involved in ocean energy development activities at present (REN21, 2010).

In principle, these technologies can supply significant amounts of energy, particularly in areas where there are large coastlines or other water features that provide many waves and large tidal movements. In fact, the theoretical potential of 7,400 EJ/year technically recoverable from the world's oceans easily exceeds current global energy requirements. The challenge remains in harnessing this potential. The IPCC (2011) finds that ocean energy technologies are unlikely to make a significant short-term contribution before 2020 due to their early stage of development.

g. Energy storage technologies

Storage technologies are critical for several RETs, especially wind and solar (PV and CSP), but also for some of the ocean technologies. The only large-scale storage technology in general use is reservoir hydro and its artificial cousin, pumped storage. A pumped storage system consists of an artificial pond or lake at the top of a hill connected to a hydraulic turbine at a lower altitude. There are quite a few pumped storage facilities in developed countries, but relatively few in developing countries, except China, due to the high front-end investment cost.

Storage technologies are critical for several RETs, especially wind and solar.

One of the technologies that can help to make storage more accessible is the "supergrid", which has been proposed by analysts to update and vastly extend the European grid. This would have the virtue of smoothing out fluctuations in local usage and in renewable supplies (wind and solar). This is because a supergrid would, for instance, be able to capture wind power over large distances (up to 500 kms away, or further, where the weather might be different), which would take care of intermittency problems as well as being able to provide electric power at a lower cost.

Denmark utilizes this approach, enabling it to export surplus wind power, when available, to neighbouring countries (including Germany, Norway and Sweden) that have hydropower which can be temporarily turned off. During the times when demand in Denmark is greater than the supply of wind, the power moves back to Denmark. Denmark's wind capacity is 3.1 GW, and it has a 1 GW connection to Norway, 0.6 GW to Sweden and 1.2 GW to Germany, or 2.8 GW altogether (Mackay, 2008). Additional R&D investment is needed to improve energy storage technologies for both wind and solar, as with some other energy technologies such as batteries for electric cars (or for providing storage with "smart" electric grids, as discussed earlier in box 2.2).

One of the technologies that can help to make storage more accessible is the "supergrid".

4. Scenarios on the future role of RETs in energy systems

Many scenarios have been developed to determine potential RET deployment rates in the future. However, it is difficult to forecast the future rate of technological progress in RETs and of the scaling up of pilot projects in various places. These will depend somewhat on future policy choices and on the scale of investments made in research, development and demonstration, as well as on deployment subsidies (discussed in chapter III). Future trends in conventional energy prices will also strongly influence the price competitiveness and deployment of RETs. In this context, the possibility that the world may be approaching a period of "peak oil" is an important consideration. Box 2.7 presents a summary of energy scenarios for the future. Under most scenarios, including those presented by the IEA (2010a), coal, oil and gas are likely to remain the predominant sources of energy until 2030 or 2035. Nuclear energy is also expected to be a significant source, but will gradually decline in importance. However, safety concerns arising in the wake of the recent nuclear reactor accident in Fukushima, Japan, may affect future trends in the use of nuclear energy.

It is forecast that RETs will continue to expand, and their share in providing renewable solutions to global energy supply will gradually rise.

Energy scenarios are sensitive to changes in their underlying assumptions, but they are useful for analytical purposes to investigate the types of energy pathways that may become possible. From the available evidence and recent trends, it is forecast that RETs will continue to expand, and their share in providing renewable solutions to global energy supply will gradually rise. They will also continue to strengthen the potential of REs as supplementary to conventional energy sources over the next two decades. Most scenarios, such as those investigated by the IEA (2010a) and IPCC (2011), show this type of outcome. It is also forecast that RETs will play a much greater role by 2030 or 2035 in diverse portfolios of energy options that

Box 2.7: Energy scenarios and the future role of RETs

Two important recent studies, one by the IPCC (2011) and the other by the IEA (2010a), have produced a number of scenarios on the future role of RETs in energy supply systems. The IPCC study considers 164 different possible scenarios on future energy mixes and climate change mitigation outcomes drawn from the literature on future energy scenarios, while the IEA study presents several possible scenarios on future energy trends. They both generally project likely growth in RETs as a growing complement to traditional energy sources, albeit with large differences in the possible roles that RETs may eventually play. It must be mentioned, however, that the "standard" scenarios used by the IPCC (developed at International Institute for Applied Systems Analysis) do not reflect the possibility that resource scarcity could limit economic growth.

The IPCC study presents potential future pathways and does not predict any particular one as being more likely than the others, though critics argue that some of them are too optimistic because they fail to take resource constraints into account. The scenarios include a wide range of possibilities regarding future climate change mitigation and the future contribution of RETs in meeting global primary energy supply. A majority of the scenarios show a substantial rise in RETs deployment by 2030, 2050 and further into the future.[a] Under these scenarios, there is no single dominant RET, but modern biomass, wind and solar generally viewed as making the largest contributions among the RETs.

The most optimistic scenario regarding the future role of RETs, not included in the IPCC study, has been prepared by Ecofys for the World Wide Fund for Nature (WWF, 2011). It suggests that 95 per cent of global energy consumption could come from RE sources by 2050. This could be achieved in part through large savings as a result of energy efficiency, so that global energy demand in 2050 is projected to be 15 per cent lower than in 2005. The IEA, by contrast, assumes continued fast economic growth and consequent growth in global energy consumption. The issue of energy efficiency is clearly vital to future trends in energy consumption. It is difficult to estimate the extent to which energy savings through efficiency gains will actually be realized, especially if they entail major lifestyle changes.

Source: UNCTAD.

[a] See UN/DESA (2011) for a discussion on the barriers to realizing the full technical potential of RE sources.

will become available to many countries. There are some scenarios which envisage RETs as playing a dominant role in the global energy system in the longer term, by 2050. Based on current trends and such projections, a recent report of the United Nations (UN/DESA, 2011) predicts that by 2050 RETs will be well on their way to completely replacing conventional energy sources worldwide. This implies extremely rapid development and deployment of RETs, combined with large energy savings through improved energy efficiency, and it would represent more of an energy revolution than the current energy evolution. The prices of energy generated by some RETs have been falling more rapidly in recent years than expected by many analysts, and RETs are playing a much bigger role today than was expected by most experts a decade ago. There could also be major technology breakthroughs in the near future as a result of ongoing R&D on various aspects of RETs, including storage and intermittency aspects, which could pave the way for a rapid acceleration in the growth of RETs. Developing broad energy matrices that promote the role of RETs in countries will help to keep options open for harnessing new technological innovations as they occur and to avoid a lock-in to specific energy technologies and infrastructures.

C. TRENDS IN GLOBAL INVESTMENTS AND COSTS OF RETS

1. Private and public sector investments in RETs

Keeping up with the trends, global investment in RETs has increased markedly during the past decade, rising from $33 billion in 2004 to $211 billion in 2010, and growing at an average annual rate of 38 per cent over that period, according to the most recent estimates available (table 2.1). This increase in investment has been closely associated with technological improvements and declining costs of RETs production, and has continued despite the global financial crisis and recession of 2008-2009 and the resulting drop in conventional energy prices. Recovery of investments was brisk in 2010 and prospects look bright for their continued growth. Nevertheless, the amount of investment remains too low to enable sufficiently rapid development and deployment of RETs.

It is predicted that by 2050, RETs will be well on their way to completely replacing conventional energy sources worldwide.

Table 2.1: Global investment in renewable energy and related technologies, 2004–2010 ($ billion)

	2004	2005	2006	2007	2008	2009	2010	Average annual growth rate 2004–2010 (%)
Investment in technology development of which:	5.3	4.7	5.7	6.7	8.2	7.6	11.0	14.6
Venture capital	0.4	0.6	1.3	1.9	2.9	1.5	2.4	46.2
Government R&D	1.1	1.2	1.3	1.5	1.6	2.4	5.3	35.1
Corporate R&D	3.8	2.9	3.1	3.3	3.7	3.7	3.3	-1.5
Investment in equipment manufacturing	0.7	4.8	14.1	25.2	19.4	15.6	18.5	139.0
Investment in RE projects of which:	26.9	47.9	72.6	109.0	140.3	141.1	193.4	41.0
Small distributed capacity	8.6	10.7	9.4	13.2	21.1	31.2	59.6	41.9
Total investment in RETs	33	57	90	129	159	160	211	38.3

Source: UNCTAD, based on UNEP and Bloomberg (2011).

Note: The data are estimates provided by Bloomberg New Energy Finance. They exclude large hydro, but include estimates for R&D investments by the private sector and governments as well as investments in small distributed RE projects.

In 2010, investment in power generation capacity accounted for 86 per cent of total RETs investment.

Government R&D investment in RETs increased significantly worldwide in both 2009 and 2010.

In the first quarter of 2009, there was a severe disruption of investments in RETs due to the crisis and the near freezing of developed-country credit markets. However, those investments recovered quickly thereafter, largely as a result of government stimulus packages in some countries, which aimed at reducing the impact of the crisis.[13] The stimulus packages offered in China and the United States are good examples of such governmental expenditure. In addition, there was continued strong investment in developing countries (UNEP and Bloomberg, 2010; World Economic Forum, 2011). In China, ambitious policy targets accompanied the financial support related to its economic stimulus plan in 2009 and 2010, mostly in wind power and PV projects. Brazil and India also continued to invest heavily in RETs (UNEP and Bloomberg, 2010; and 2011).

In 2010, investment in power generation capacity accounted for 86 per cent of total RETs investment (or $181 billion). Most of this investment was in large, utility scale projects (mainly large wind farms, solar parks, biomass power plants and biofuel refineries). Investment in small-scale renewable energy projects (mostly in rooftop solar PV panels) has also been rising rapidly since 2009, along with increased activity in small distributed capacity, partly stimulated by dramatic price declines of solar PV modules and systems (UNEP and Bloomberg, 2011; REN21, 2011). In addition, investments of $40–$45 billion were made in large hydro plants in 2010, and estimated investments of about $15 billion in solar hot water collectors, neither of which are reflected in table 2.2 (REN21, 2011). If investments in solar hot water collectors are included, the global total in 2010 increases to $226 billion (excluding the $40 billion invested in large hydro projects).

Investment in RETs aimed at greater renewable capacity has three main financing components: (i) asset financing of utility-scale projects, (ii) refinancing and acquisition of such projects, and (iii) financing of small-scale projects. In 2010, by far the largest amount of asset financing went to large utility-scale RE projects (86 per cent of the total), and a very small share targeted equipment manufacturing (8.8 per cent) and technology development (5.2 per cent). Investment in technology development included both government and private R&D investment and venture capital financing (UNEP and Bloomberg, 2011).

Government R&D investment in RETs increased significantly worldwide in both 2009 and 2010, in keeping with the overall trend observed between 2004 and 2010 when such investments rose at an annual average rate of 35 per cent (table 2.1). Financing by State-owned multilateral and bilateral development banks rose following distress in private capital markets in 2008. Globally, 13 development banks provided financing of $13.5 billion for RETs projects in 2010 – a marked increase over their financing in 2009 ($8.9 billion), 2008 ($11 billion) and 2007 ($4.5 billion) (UNEP and Bloomberg, 2011). Public policy support, including direct government support and development bank financing, has played an important role in maintaining the rate of RETs development and deployment since 2008. This is in marked contrast to the stagnation in corporate R&D in RETs, which declined by an average annual rate of 1.5 per cent during the period 2004–2010.

Available data indicate that wind power has been by far the largest recipient since 2007, with new financial investments of $95 billion in 2010,[14] representing 66 per cent of the total. Solar has been the second largest with $26 billion (18 per cent), followed by modern biomass power with $11 billion (8 per cent) and biofuels with $6 billion (4 per cent). As noted earlier, $40–$45 billion were invested in large hydro projects (not included in the data), whereas investments in small hydro, geothermal and marine energy were much smaller (UNEP and Bloomberg, 2011).

Foreign direct investment (FDI) in RETs (including electricity generation and the manufacturing of RETs equipment) grew rapidly over the period 2003–2010, at an average annual rate of 43.4 per cent. Growth was especially rapid in 2006, 2007 and 2008, after which the dislocation in international credit markets associated with the global financial crisis led to declines in FDI flows in 2009 and again in 2010. Developed economies were the main investors, accounting for 89.5 per cent of all FDI in renewable energy over the period 2003–2010. They were also the largest hosts of investments in RETs, accounting for 55 per cent of global FDI inflows in RETs. Developing economies accounted for 9.6 per cent of global RETs-related FDI outflows, and they invested mainly in other developing economies (see data and discussion in chapter V, box 5.14).

Higher oil prices, renewed concern over the safety of nuclear energy and continued public policy support should provide the basis for continued private and public investment in RETs. Against this positive picture, private capital markets and private banks have still not fully recovered from the global financial crisis that erupted in 2008. International debt markets remain unsettled, and private investors in many developed countries continue to remain constrained by reduced access to financing from banks and international capital markets. In addition, a number of developed countries are pursuing policies of fiscal austerity, which may cause them to reconsider their direct fiscal support measures for renewable energy,[15] although other support measures (such as RE targets, renewable portfolio standards, and/or tax credits) may remain in place. The picture is therefore mixed.

However, for purposes of this Report, the important issue is whether public support measures and investment will continue to provide the much-needed impetus to RETs innovation and its wider dissemination over the long term. Will investment in renewable energy rise at rates that are fast enough to meet the global challenge of reducing energy poverty and mitigating climate change? Viewing the current trends and data available from this perspective, much more private investment will be needed to accelerate the development and deployment of RETs worldwide. It has been estimated, for instance, that at least $500 billion will need to be invested in new, low-carbon technologies each year starting in 2020 in order to stabilize climate change (BNEF, 2010: 1). This applies particularly to many developing countries and LDCs where investments in RETs have been much lower than in some of the larger developing countries.[16]

Higher oil prices...and continued public policy support should provide the basis for continued private and public investment in RETs.

From a different perspective, large investments are also needed in order to scale up RETs production and reduce unit costs via the "experience curve" – which reflects economies of scale and "learning by doing". This is very important to enable wider deployment in several smaller developing countries and LDCs that may not be able to provide direct financial support for RETs.

Financing for improved RETs infrastructure, transmission and distribution is particularly important to enable the greater deployment of some RETs. This will require much greater investments, including for upgrading the energy network infrastructure (including "smart grids"), to enable faster deployment of RETs so as to mitigate climate change and reduce energy poverty.

2. Costs of renewable energy and other energy sources compared

The cost of RE-generated electric power will have a major impact on the extent of its use among poor households in developing countries. Even in developed countries, large-scale deployment of REs rests on their ability to compete with conventional fossil fuels in terms of price. Comparing costs of REs with conventional energy sources is difficult because certain costs are specific to conventional energy and others to REs, and they are difficult to factor into any financial equation, as the earlier discussions in this chapter show. This is compounded by the fact that certain large-scale applications of REs are subsidized through fiscal support by governments in many developed countries in order to compensate for conventional energy sources that are not sold at their true price. These issues are exam-

Financing for improved RETs infrastructure, transmission and distribution is particularly important to enable the greater deployment of some RETs.

ined here, followed by a discussion of the difficulties in incorporating the true price of energy into market rates.

Public fiscal support... is much easier to justify on grounds that RETs will render such energy options more price-competitive over time.

a. Problems with making direct cost comparisons

(i) Fiscal support by governments

Currently, public fiscal support plays a role in ensuring the price competitiveness of RE in many developed countries, and it is feared that such support may eventually be withdrawn under the pressure of fiscal austerity aimed at alleviating the heavy debt burdens of those countries. Such support is much easier to justify on grounds that RETs will render such energy options more price-competitive over time, and because of the other benefits they can offer, such as reduction of energy poverty, climate change mitigation, job creation and poverty reduction (see chapter III). Rising oil prices and increasing financial investments and speculation in energy commodities are also factors that promote the greater use of RETs for energy security purposes. And an accelerated use of RETs would in turn result in further cost reductions (see, for example, IEA, 2010a and discussion in chapter III of this Report). Moreover, the price differential will decrease as fossil fuel prices rise as a result of increasing global energy demand in the coming decades.

(ii) Factoring in costs specific to conventional energy: Subsidies and environmental externalities

In addition to the hidden subsidy from not costing the environmental impact of fossil fuels, many countries have been providing direct subsidies to fossil fuel consumption.

One clear problem in calculating the true cost of energy lies in the difficulty of accounting for the externalities inherent in utilizing fossil fuels. Fossil fuel combustion causes a number of harmful emissions, including micro-particulates, sulfur and nitrogen oxides, volatile hydrocarbons and GHGs, which contribute to climate change. The fact that the costs resulting from these emissions are not taken into account when calculating the overall cost of fossil fuel constitutes a large, hidden subsidy. Unfortunately, the true magnitude of that subsidy is difficult to quantify with any precision because of the number of assumptions that need to be made with regard to the environmental impact of fossil fuels. For example, nuclear energy – which benefits from limited liability laws in many countries – creates large public costs when accidents occur such as at Chernobyl and Fukushima. This is in addition to the unresolved problem of storage of nuclear waste and the costs of decommissioning nuclear plants. The risk of nuclear accidents is incalculable because of the human element, and this presents difficulties in assigning probabilities or prices to that risk.

In addition to the hidden subsidy from not costing the environmental impact of fossil fuels, many countries have been providing direct subsidies to fossil fuel consumption for many years. A common reason for such subsidies in some developing countries is to protect poor households from rising energy prices, or to promote access to modern energy sources by the poor. Subsidies have also been used to promote the development, deployment and use of nuclear energy (IEA et al, 2010).

Energy subsidies can take many forms, both direct and indirect. Direct subsidies may be fairly easy to account for. However, indirect ones, including tax exemptions and preferential tax rates (e.g. reduced value-added tax (VAT) rates or exemptions from excise duties for fossil-fuel use) are difficult to measure, but may also exist.[17] Governments also provide subsidies for some RETs to encourage their development and deployment, but these are much smaller. It is estimated that fossil fuel subsidies to consumers amounted to $557 billion in 2008 (IEA, 2010a). Producing comprehensive estimates are more difficult for nuclear energy and RETs, but a rough estimate of $100 billion annually for alternative energies (including both nuclear and RETs) has been reported (IEA et al., 2010).

(iii) Factoring in costs specific to RETs

Intermittency in power generation is a feature of most RETs, and this has implications for efficiency and cost of electricity generation. A more general problem is accounting for energy costs on a life-cycle basis, which includes all the costs involved in an energy project, from start to end. Life-cycle cost accounting, which is a more comprehensive method of calculating costs, can be used for comparisons of different energy sources. It includes project investment costs, operation and maintenance costs, and decommissioning costs once the useful life of a mine, a drilling rig or a power plant has ended. Moreover, the costs of renewable energy are likely to vary further depending, for example, on the geographic location and the natural resource endowments needed to generate power. The issue of storage and transmission infrastructure is also relevant in this context, because the distances involved in collecting and transporting solid fuels, such as biomass from agricultural waste, from the point of generation to where it is processed vary by location. This is apart from standard considerations such as the cost of land or of labour to install some RETs.

The intermittency of some RETs poses a number of challenges relating to the need for reserve generation over and above responding to increases in demand or plant failure. One challenge is that capacity has to be held in reserve to deal with short-term fluctuations in RETs-based output. For instance, when the wind is too low or too strong for a wind farm to operate, reserve capacity has to be brought online. Another challenge associated with intermittency arises from the need to hold excess capacity (so-called "capacity credit") in order to meet peak demand.

Like conventional plants, intermittent RETs suffer from risks of technical failure, but they also suffer from the risk that their "fuel" may not be available. And furthermore, if the "fuel" is not available at one plant, it is highly unlikely to be available at nearby plants. Overcoming these intermittency challenges incurs costs associated with the need for back-up supply. From a purely financial perspective, the value of generation from intermittent RETs should be lower than that from conventional energy plants by roughly the amount of these additional back-up costs in order to make RETs competitive (Owen, 2004).

b. Incorporating costs into the market price of energy options

The ideal approach to incorporating environmental costs into market pricing is based on the 'polluter pays' principle, but this is seldom enforced. Apart from uncertainties about the externality costs (noted above), there are serious implementation difficulties to be overcome. For energy technologies relating to climate change mitigation, this might be done through carbon pricing or some form of exchangeable emission rights (discussed in chapter IV). By regulating a particular price for one ton of CO_2-equivalent emissions, project developers and investors are forced to include this cost when deciding on a particular technology. For instance, GHG-emitting fossil-fuel power plants would become more expensive because the cost-benefit analysis of building and generating power from them would not only include the costs associated with capital investment, operation and maintenance, and fuel, but also the cost of GHGs emitted based on the carbon price. Other environmental impacts, such as localized air pollution, deforestation from unsustainable biomass use, loss of biodiversity from deforestation and/or pollution, are not necessarily accounted for in a carbon price, although adjustments are possible. Other measures, which avoid the difficulties of pricing externalities, may include mandating regulated electricity suppliers to provide a certain proportion of their electricity from "green" sources.

Intermittency in power generation is a feature of most RETs, and this has implications for efficiency and cost of electricity generation.

In the absence of a market price for carbon (or other GHG emissions), subsidies for RETs may serve to compensate for some externalities. They are also easier to justify on theoretical grounds, because of their environmental, health and social benefits. In addition, the many market failures in technology markets can justify public intervention to promote technology development in technologies that have high social welfare returns, and that bring various social and

In the absence of a market price for carbon (or other GHG emissions), subsidies for RETs may serve to compensate for some externalities.

Most studies on RE costs agree that the costs of energy generation from various RETs are on the decline, and that this trend will continue over time.

Major technological advances and associated cost reductions are expected in, for instance advanced PV and CSP technologies and manufacturing processes.

economic benefits that cannot be directly accounted for in their price.[18] For example, providing basic energy services to energy-poor communities can be seen as desirable, or even as an obligation of a government to meet the most basic needs of its citizens. And actions that reduce GHG emissions and mitigate climate change are considered global public goods. Thus, there are economic arguments to support such actions through policy measures, although the cost of such support remains a valid issue. An argument in support of RETs use and deployment is that many RETs are still not mature, and their costs are likely to fall through economies of scale and with experience. Of course, this dynamic aspect cannot be fully factored into current comparisons, but it needs to be borne in mind.

3. The evidence on renewable energy costs

At the microeconomic or project level, power projects are most commonly appraised on the basis of their levelized cost of electricity (LCOE).[19] Assuming the project operates at full capacity, the LCOE is determined by comparing the discounted capital cost of the project, the annual operating and maintenance costs and the expected annual fuel costs, on the one hand, with the expected annual production of electricity on the other (Heal, 2009; Owen, 2004). It therefore takes into account all the financial costs involved in a project (investment, operation and maintenance, fuel and decommissioning costs) and amortizes these costs over the expected life of a project. Usually, LCOE calculations do not take into account subsidies or policy incentives for RETs.

Different studies on costs of REs lead to different conclusions on their cost competitiveness in relation to other sources of energy. One reason for this, as discussed earlier in this chapter, is that RE resources and costs differ substantially by location and by project. The assumptions made regarding discount rates can also have an important impact on the LCOE. The relative economics of RE versus conventional sources is largely driven by forecasts of fuel prices and future technology costs and performance, together with the prices of certain construction and manufacturing materials such as steel, concrete, glass and silicon (ESMAP, 2007). As discussed in this section, evidence indicates that for some applications RETs are already cost-competitive compared with traditional energy sources.

Most studies on RE costs agree that the costs of energy generation from various RETs are on the decline, and that this trend will continue over time. Major technological advances and associated cost reductions are expected in, for instance advanced PV and CSP technologies and manufacturing processes, enhanced geothermal systems, multiple emerging ocean technologies, and foundation and turbine designs for offshore wind energy. Further cost reductions in hydropower are likely to be less significant than in some of the other RE technologies, but there is potential for R&D to make hydropower projects technically feasible in a wider range of natural conditions, and to improve the technical performance of new and existing projects (IPCC, 2011).

One study by ExxonMobil (2010) on global energy trends to 2030 forecasts that coal, gas and nuclear energy will remain more price-competitive than solar PV and geothermal for new, base-load power-generation plants that come online in the United States in 2025, although by then, wind will have become more competitive than all three (coal, gas and nuclear). Wind is more competitive than, and geothermal is as competitive as, coal that incorporates carbon capture and storage (which is more costly than "dirty" raw coal). The study's projections indicate that both wind and solar will become much larger sources of power generation in coming decades. It should be noted that the price projections may differ for power plants in different countries, given that there are large location- and project-specific variations. The study does not appear to take into account different configurations and scales of application of RETs. Moreover, the assumptions underlying the study are not explicitly noted.

Taking account of different configurations and scales of application provides more nuanced results. This is illustrated by the results of a detailed study, which found that in some off-grid and mini-grid applications certain RETs were already competitive with conventional energy in 2005, even with the relatively low oil prices prevailing at that time (ESMAP, 2007). This implies that for precisely those applications which may be most suitable for isolated communities (i.e. decentralized applications that do not require connection to the national or regional energy grids) RETs may be at their most cost-competitive. Moreover, it is worth noting that cost reductions have been more rapid in some REs than the study had foreseen, notably in solar PV systems. These cost reductions are an important trend for many people in developing countries who suffer from energy poverty and for whom the most relevant consideration is the price of RETs-based energy supply.

It is clear that actual energy generation costs (in terms of the localized cost of electricity) based on some RETs have been declining over time, and in some cases very rapidly. The prices of solar-PV systems, in particular, have been falling extremely rapidly, by a factor of 10 for PV modules over the past 30 years (from $22/W in 1980 to less than $1.50/W in 2010). The price of an entire system has also declined steadily, reaching $2.72/W for some thin-film technologies by 2009 (IPCC, 2011). During the 18 months to June 2010, prices fell by an estimated 50 per cent for new solar panel modules (BNEF, 2010: 4).

Between 2008 and 2009, the LCOE range for thin-film PV completely shifted, and traditional crystalline silicon PV modules became much cheaper (figure2.3). The cost of thin-film PV reported here appears to be lower than that projected by ESMAP (2007) for off-grid and mini-grid applications in 2015, since actual progress is much faster than was projected. It is reported that in Africa, Asia and Latin America, the demand for modern energy is driving the use of PV for mini-grid or off-grid solar systems, which in many instances are already at price parity with fossil fuels (REN21, 2010). Such systems are contributing significantly to reducing energy poverty by providing access to energy where grid connection remains elusive. Some other recent findings on the declining costs of RETs are presented in box 2.8 below.

Between 2008 and 2009, the LCOE range for thin-film PV completely shifted, and traditional crystalline silicon PV modules became much cheaper.

Figure 2.3: Levelized costs of some renewable energy technologies compared, 2008 and 2009 ($/MWh)

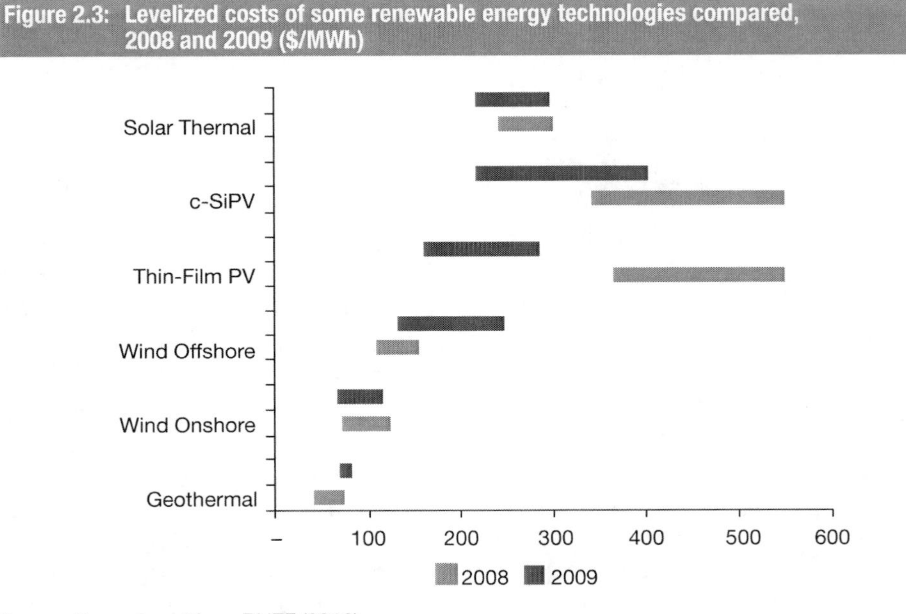

Source: Reproduced from BNEF (2010).

Note: c-Si refers to traditional crystalline silicon PV modules. The above data are based on an assumed expected internal rate of return of 10 per cent for investors in such generating projects.

Box 2.8: Declining costs of RETs: Summary of findings of the IPCC

The IPCC (2011) has reviewed a broad number of studies which provide additional support for the view that some RETs are already cost-competitive under some circumstances. The main findings for solar, wind, hydropower and geothermal are discussed below.

The localized cost of *PV* depends heavily on the cost of individual system components, of which the PV module is the most costly. The current LCOE from solar PV is generally still higher than wholesale market prices for electricity, although in some applications PV systems are already competitive with other local conventional energy alternatives for energy access. Recent LCOEs for different types of PV systems showed wide variations, from as low as $0.074/kWh to as high as $0.92/kWh, depending on a wide set and range of input parameters. Narrowing the range of parameter variations, the LCOE in 2009 for utility-scale PV electricity generation in regions of high solar irradiance in Europe and the United States was reported to be in the range of $0.15/kWh to $0.4/kWh at a 7 per cent discount rate, but it could be lower or higher depending on the available resource and other conditions. These calculations on the LCOE for different RETs show that the cost of solar PV energy remains higher than that for other RETs, but this may not fully take into account the recent rapid decline in the price of PV technologies.

For *onshore and offshore wind*, the LCOE varies substantially, depending on assumed investment costs, energy production and discount rates. In some areas with good wind resources, the cost of wind energy is already competitive with current energy market prices, even without considering externalities of conventional energy such as environmental impacts. For onshore wind energy in good to excellent wind resource regimes, the IPCC estimates the LCOE to average 5 cents/kWh to 10 cents/kWh, and it could reach more than 15 cents/kWh in areas with poor wind conditions. Although the offshore cost estimates are less certain, typical LCOEs are estimated to range from 10 cents/kWh to more than 20 cents/kWh for recently built or planned plants located in relatively shallow water.

Hydropower is often economically competitive with traditional energy, although the cost of developing, deploying and operating new hydropower projects varies from project to project. This is because hydro projects differ greatly in nature. The LCOE of hydropower projects, using a large set and range of input parameters, ranges from as low as 1.1 cents/kWh to 15 cents/kWh, depending on site-specific parameters for investment costs of each project and on assumptions regarding the discount rate, capacity factor, lifetime, and operating and maintenance costs. Under favourable conditions, where the costs of those parameters are low, the LCOE of hydropower can be in the range of 3 cents/kWh to 5 cents/kWh.

Geothermal costs also vary by project, but the LCOEs of power plants using hydro-thermal resources are reportedly often competitive in electricity markets. The same is reported to be true for direct uses of geothermal heat. The LCOE of geothermal projects based on a large set and range of input parameters is estimated to range from 3.1 cents/kWh to 17 cents/kWh, depending on the particular type of technology and project-specific conditions.

Source: UNCTAD.

RETs can also be particularly competitive for heating and cooling.

Table 2.2 shows the costs of energy produced by various RETs. The table helps to underscore two important points. First, the energy costs generated by RETs vary substantially by application, as noted earlier in this chapter. Second, for some applications, RE costs vary within a relatively narrow range, for instance: 14 cents/kWh–18 cents/kWh for energy from a CSP power plant, 17 cents/kWh–34 cents/kWh from a rooftop solar PV, and 5 cents/kWh–12 cents/kWh either from a small biomass power plant of 1 to 20 MW, or a mini or small hydro installation. RETs can also be particularly competitive for heating and cooling. The price ranges shown in table 2.2 can be very broad due to location- and project-specific variations. Individual projects falling at the lower end of LCOE ranges are relatively cost-competitive.

Recent data from the IEA (2010b) allow a comparison of the LCOE of conventional sources of energy with those of onshore wind and solar PV systems. The data present the median values (in $/MWh) of LCOE cost ranges based on data collected on different energy installations in various countries using discount rates of 5 per cent and 10 percent respectively. The results indicate that with a discount rate of 5 per cent, nuclear power plants generate the cheapest electricity among the technologies studied, at $59/MWh, followed closely by coal-fired plants ($62/MWh for coal plants with

Table 2.2: RETs characteristics and energy costs

Technology	Typical characteristics	Typical energy costs (cents/kWh, unless indicated otherwise)
Power Generation		
Large hydro	Plant size: 10 MW–18,000 MW	3–5
Small hydro	Plant size: 1–10 MW	5–12
Onshore wind	Turbine size: 1.5–3.5 MW Blade diameter: 60–100 meters	5–9
Offshore wind	Turbine size: 1.5–5 MW Blade diameter: 70–125 meters	10–20
Biomass power	Plant size: 1–20 MW	5–12
Geothermal power	Plant size: 1–100 MW; Types: binary, single- and double-flash, natural steam	4–7
Solar PV (module)	Cell type and efficiency: crystalline 12–19%; thin film 4–13%	—
Rooftop solar PV	Peak capacity: 2–5 kW-peak	17–34
Utility-scale solar PV	Peak capacity: 200 kW–100 MW	15–30
Concentrating solar thermal power (CSP)	Plant size: 50–500 MW (trough), 10–20 MW (tower); Types: trough, tower, dish	14–18 (trough)
Hot Water/Heating/Cooling		
Biomass heat	Plant size: 1–20 MW	1–6
Solar hot water/heating	Size: 2–5 m^2 (household); 20–200 m^2 (medium/multi-family); 0.5–2 MWh (large/district heating); Types: evacuated tube, flat–plate	2–20 (household) 1–15 (medium) 1–8 (large)
Geothermal heating/cooling	Plant capacity: 1–10 MW;	0.5–2
Biofuels		
Ethanol	Feedstocks: sugar cane, sugar beets, corn, cassava, sorghum, wheat (and cellulose in the future)	30–50 cents/liter (sugar) 60–80 cents/liter (corn, gasoline equivalent)
Biodiesel	Feedstocks: soy, rapeseed, mustard seed, palm, jatropha, and waste vegetable oils	40–80 cents/liter (diesel equivalent)
Rural Energy		
Mini-hydro	Plant capacity: 100 –1,000 kW	5–12
Micro-hydro	Plant capacity: 1–100 kW	7–30
Pico-hydro	Plant capacity: 0.1–1 kW	20–40
Biogas digester	Digester size: 6–8 m^3	n/a
Biomass gasifier	Size: 20–5,000 kW	8 –12
Small wind turbine	Turbine size: 3–100 kW	15–25
Household wind turbine	Turbine size: 0.1–3 kW	15–35
Village-scale mini-grid	System size: 10–1,000 kW	25–100
Solar home system	System size: 20–100 W	40–60

Source: Reproduced from REN21 (2011: 33).

CCS and $65/MWh for supercritical/ultra-supercritical coal-fired plants). Electricity from combined cycle gas turbine (CCGT) plants is more expensive, at $86/MWh, onshore wind is even more expensive ($96.74/MWh) and solar PV is by far the most costly of the group (at $411/MWh). With a 10 per cent discount rate, the order of competitiveness shifts for coal, gas and nuclear, but onshore wind and solar PV remain the most expensive by a substantial margin. Based on the data reported here, on average, onshore wind and solar PV seem to remain more costly than conventional energy sources.

In small-scale applications that are off-grid or integrated into mini-grids, RETs are already competitive, providing solutions that are difficult for conventional grid-based energy sources.

In summary, there is evidence that while in many large-scale applications conventional energy is often more cost-competitive, this is not always the case. Some RETs are cost-competitive, although cost characteristics are highly context-specific and can vary from one system to another. In small-scale applications that are off-grid or integrated into mini-grids, RETs are already competitive, providing solutions that are difficult for conventional grid-based energy sources to emulate. In that respect, it is reasonable to conclude that the wider deployment of RETs could make an important contribution to reducing energy poverty, particularly through off-grid and mini-grid applications. These smaller, decentralized applications should be particularly useful in small isolated communities in LDCs and other developing countries, as well as in developed countries, assuming that the required RE resources are available in those locations. It must be borne in mind that some of the RETs remain dynamic and subject to ongoing technological development. As they are further scaled up, and with growing experience and technological learning, prices can be expected to drop further.

D. SUMMARY

The basic messages of this chapter may be summarized in the following five points.

First, RETs are a diverse group of technologies that are currently at different levels of maturity. Those based on wind, geothermal, solar thermal and hydro are mature technologies and are already being deployed widely. Solar concentrators and solar PV systems are rapidly penetrating new markets, even as development continues. Still others, including second-generation biofuels and ocean energy, remain at varying stages of pre-commercial development.

Second, some RETs for off-grid and mini-grid applications may also already provide cost-effective energy solutions. There have been rapid cost reductions in solar PV, but the relative cost competitiveness of different PV technologies is not clear. Still, where good alternatives do not exist, solar PV can represent a reasonable option to provide some degree of access to energy, particularly in rural areas in developing countries and LDCs where national energy grids are unlikely to expand in the near future. In these cases, RETs offer a realistic option for eradication, or at least for alleviation, of energy poverty. In some developing countries that lack adequate physical infrastructure, grid connection rates are extremely low and RETs could provide alternate energy supply sources for large segments of the population.

Third, some RETs are experiencing rapid ongoing technological progress and reductions in energy generation costs, particularly of solar PV technologies, but also of onshore wind energy. The cost competitiveness of RETs relative to conventional energy sources is improving, and can be expected to improve even further with continued technological progress and higher investment in development, production and deployment. Rising, and increasingly volatile, oil prices may also be contributing to this trend. Additional technological progress is needed for integrating RE into the existing energy infrastructure, including through the development of smart energy grids. Such grids could help overcome the intermittency problem associated with energy from solar and wind energies. Also, further progress in the storage capabilities for these two RETs is needed.

Fourth, some RETs are being deployed rapidly, but are starting from a small base, and therefore still account for only a small fraction of global energy consumption today. However, the rate of growth of global investment and deployment in RETs has risen over the past decade. Both developed and developing countries are participating in this growth, although there is a need for expanding RETs-related investment in smaller developing countries and LDCs.

Fifth, there is huge technical potential for power generation from RETs, and therefore they are likely to play an increasingly important role in meeting global energy demand as continued technological progress, additional investment and further deployment lead to cost reductions over the medium and long term globally.

This chapter has presented and assessed the many different scenarios on the future role that RETs could play in global energy supply. The analysis shows that RETs will continue to evolve as complements to existing energy sources globally, with the eventual aim of replacing conventional energy in the long term. For developing countries and LDCs, this is a positive trend. The actual speed and extent of deployment of RETs and the role they will eventually play will depend critically on the policy choices that are made today and in the future. The policy issues that need to be considered within national frameworks for technology and innovation and the ways and means of international support are discussed in subsequent chapters of this Report.

NOTES

1 This definition has been in use since the 1980s (see, for example, Twidell and Weir, 1986).

2 The use of bagasse for power and heat production (for example, through cogeneration) is reportedly significant in developed and developing countries that have a large sugarcane industry (REN21, 2011).

3 The IPCC notes the same point for the year 2008, when the technical potential of renewable energy to generate electricity was much greater than global electricity demand registered that year (IPCC, 2011).

4 The Middle East figures referred to here correspond to Global Wind Energy Council (GWEC) classification.

5 The cost of a fossil fuel generator will depend upon the type of generator (soundproof or not) as well as the type of fuel it runs on.

6 According to standard terminology, pico-hydro has the smallest power capacity, of 0.1–1 kW, micro-hydro covers a range of 1–100 kW, mini-hydro a range of 100–1,000 kW (1 MW) and small hydro a range of 1–10 MW, while large hydro implies capacity of over 10 MW. Hydropower plants can be classified into three main categories according to operation and type of flow: run-of- river, reservoir based (storage) hydropower and pumped storage. Each of these has different variability and predictability characteristics with respect to power generation.

7 The IPCC (2011) estimates the technical potential of global hydropower to be 14,575 TWh (or 52.47 exajoules (EJ)).

8 The resulting gas consists of 50–70 per cent methane, 5–10 per cent hydrogen, and the rest mainly CO_2 (FAO, 2006). This anaerobic decay happens naturally under silt in swamps, but the process can be simulated using fairly simple equipment, the main purpose of which is to keep air away from the waste materials and to capture the methane as it is released.

9 Neodymium is one of the "heavy" rare earths, currently produced almost exclusively in China.

10 The solar radiation itself has a temperature of nearly 5,777 degrees Kelvin, or 5,477 degrees Celsius, so the receiver gets very hot and transmits that heat into a "working fluid", which could be helium or water or one of the hydrocarbons. However, the mirrors on the ground have to be controllable to keep aimed at the sun as it moves across the sky.

11 For more information, see: www.desertec-africa.org.

12 Many such facilities have been built in Japan, Spain and the United States. Costs per kWh are still much higher than for coal-based electricity, but solar PV costs continue to decline, especially since China recently began exporting solar panels on a large scale. Several PV technologies have been developed in parallel. The panels can be arranged on virtually any scale, from a single panel on a rooftop to multiple panels organized in arrays to form solar farms.

13 Under those packages, over $194 billion were part of government commitments to spending on clean energy (World Economic Forum, 2011).

14 Financial investment refers to asset finance, plus capital raising by companies from venture capital, private equity and public market investors. Excludes small-scale projects and government and corporate R&D.

15 These include cuts in national and State/province-level support in 2010 in the Czech Republic, France, Italy, Germany, Spain and the United Kingdom (REN21, 2010).

16 Larger developing countries, especially China, are beginning to receive the largest share of new investments in RETs (UNEP and Bloomberg, 2011).

17 Examples include reduced VAT on fossil fuels in Italy and the United Kingdom, and variable excise taxes on petrol and diesel in Mexico (for other examples, see IEA et al., 2010).

18 See, for example, UNCTAD (2010) on RETs-based electrification in rural areas.

19 It should be noted that the use of LCOEs to compare intermittent REs (such as wind or solar) with dispatchable energy sources has been criticized as being faulty because the wholesale price of electricity varies widely over a day, month or year, and because intermittent generating technologies have very different energy production profiles than dispatchable ones (Joskow, 2011). Joskow proposes that such comparisons utilize evaluations based on three elements: the expected market value of the electricity that will be supplied, total life-cycle costs and expected profitability.

REFERENCES

Aitken JW (2003). Transitioning to a renewable energy future. Freiburg, International Solar Energy Society.

BNEF (Bloomberg New Energy Finance) (2010). Crossing the valley of death: Solutions to the next generation energy problem. June. Available at: http://www.cleanegroup.org/assets/Uploads/CEGBNEF-2010-06-21valleyofdeath.pdf.

Boyle G (2004). *Renewable Energy: Power for a Sustainable Future*. Second edition. Glasgow, Oxford University Press.

Briggs M (2004). Widescale biodiesel production from algae. *Energy Bulletin,* October. Durham, NH, University of New Hampshire.

Elswijk M and Kaan H (2008). European embedding of passive houses. Promotion of European Passive Houses. Available at: http://erg.ucd.ie/pep/pdf/European_Embedding_of_Passive_Houses.pdf

ESMAP (2007). Technical and economic assessment of off-grid, mini-grid and grid electrification technologies. Technical Paper no. 121/07, Washington, DC, World Bank, December.

Eyer J and Corey G (2010). Energy storage for the electricity grid: Benefits and market potential assessment guide no. SAND2010-0815. Alberquerque, NM, Sandia National Laboratories.

ExxonMobil (2010). The outlook for energy: A view to 2030. Available at: http://www.exxonmobil.com/Corporate/files/news_pub_eo.pdf.

FAO (2006). A system approach to biogas technology. Rome.

Ford Runge C and Senauer B (2007). How biofuels could starve the poor. *Foreign Affairs*, May/June Tampa, FL, Council on Foreign Relations Inc.

GWEC (2011). Annual market update 2010, March.

Greenpeace International, SolarPACES and ESTELA (2009). *Concentrating Solar Power – Global Outlook 09: Why Renewable Energy is Hot*. Available at: http://www.greenpeace.org/raw/content/international/press/reports/concentrating-solar-power-2009.pdf

Heal G (2009). The economics of renewable energy. Cambridge, MA, National Bureau of Economic Research, June.

IEA (2007). *Renewables in Global Energy Supply 2007*. Paris.

IEA (2010a). *World Energy Outlook 2010*. Paris, OECD/IEA.

IEA (2010b). *Key World Energy Statistics 2010*. Paris.

IEA, OPEC, OECD and World Bank (2010). Analysis of the scope of energy subsidies and suggestions for the G-20 Initiative, IEA, OPEC, OECD, WORLD BANK Joint Report, June.

IPCC (2011). *IPCC Special Report on Renewable Energy Sources and Climate Change Mitigation*. Geneva.

Johansson TB (2011). Development of sustainable energy. Paper presented at the 19th OSCE Economic and Environmental Forum first preparatory meeting in Vienna, 7 February 2011. Available at: http://www.osce.org/eea/75430.

Joskow P (2011). Comparing the costs of intermittent and dispatchable electricity generating technologies. *American Economic Review,* 101(3): 238–241, May. American Economic Association.

MacKay DJ (2008). *Sustainable Energy - Without the Hot Air*. Cambridge, UIT Cambridge Ltd.

Owen A (2004). Environmental externalities, market distortions and the economics of renewable energy technologies. *The Energy Journal*, 25(3): 127–156.

Pew Charitable Trusts (2011). Who's winning the clean energy race? Washington, DC. Available at: www.pewenvironment.org/uploadedFiles/PEG/Publications/Report/G-20Report-LOWRes-FINAL.pdf

Photon International (2011). Global PV cell production expanded 118% to 27.2 GW in 2010. *SolarServer.com solar magazine*, 18 January.

Pollin R, Heintz J and Garrett-Peltier H (2009). The economic benefits of investing in clean energy. How the economic stimulus program and new legislation can boost U.S. economic growth and employment. Amherst, MA, Department of Economics and Political Economy Research Institute (PERI), University of Massachusetts.

REN21 (2005). *Renewables 2005: Global Status Report*. Washington, DC, Worldwatch Institute.

REN21 (2010). *Renewables 2010: Global Status Report*. Paris, REN21 Secretariat.

REN21 (2011). *Renewables 2011: Global Status Report*. Paris, REN21 Secretariat.

Singer P (2010). Energy storage and the grid. *Renewable Energy World North America Magazine*. December. Available at: http://www.renewableenergyworld.com/rea/news/article/2010/12/energy-storage-and-the-grid

Sorensen, B (2000). *Renewable Energy* (Second Edition). Academic Press, London.

Twidell J and Weir D (1986). *Renewable Energy Resources*. London, E & FN Spon.

UNCTAD (2008). *Biofuel Production Technologies: Status, Prospects and Implications for Trade and Development*. New York and Geneva, UNCTAD.

UNCTAD (2010). *Renewable Energy Technologies for Rural Development*. New York and Geneva, UNCTAD.

UN/DESA (2011). *World Economic and Social Survey 2011*: *The Great Green Technological Transformation*. New York, United Nations Department of Economic and Social Affairs.

UNEP and Bloomberg (2010). *Global Trends in Sustainable Energy Investment 2010: Analysis of Trends and Issues in the Financing of Renewable Energy and Energy Efficiency.* United Nations Environment Programme. Available at: http://bnef.com/Download/UserFiles_File_WhitePapers/sefi_unep_global_trends_2010.pdf.

UNEP and Bloomberg (2011). *Global Trends in Renewable Energy Investment 2011. Analysis of Trends and Issues in the Financing of Renewable Energy.* United Nations Environment Programme. Available at: www.fs-unep-centre.org/publications/global-trends-renewable-energy-investment-2011.

UNEP, European Patent Office and ICTSD (2010). *Patents and Clean Energy: Bridging the Gap between Evidence and Policy*. Final report. Available at: http://documents.epo.org/projects/babylon/eponet.nsf/0/cc5da4b168363477c12577ad00547289/$FILE/patents_clean_energy_study_en.pdf.

WEC (2010). *2010 Survey of Energy Resources*. London.

World Economic Forum (2011). *Green Investing 2011: Reducing the Cost of Financing*. Geneva, April.

WWF (World Wildlife Fund) (2011). The energy report: 100% renewable energy by 2050. Available at: http://wwf.panda.org/what_we_do/footprint/climate_carbon_energy/energy_solutions/renewable_energy/sustainable_energy_report/.

3 STIMULATING TECHNICAL CHANGE AND INNOVATION IN AND THROUGH RENEWABLE ENERGY TECHNOLOGIES

CHAPTER III

STIMULATING TECHNICAL CHANGE AND INNOVATION IN AND THROUGH RENEWABLE ENERGY TECHNOLOGIES

A. INTRODUCTION

The historically unprecedented rapid growth rates in the industrialized countries over the past century have been made possible by the application of science to productive activities, which underscores the importance of knowledge and innovation for growth, competitiveness and poverty reduction (Rosenberg, 1982; Baumol, 2003; Nelson and Winter, 1982; Reinert, 2007). This growth has led to a widening gap between the industrialized and developing countries; but also over the past three decades, there has been a growing divergence among developing countries themselves (Ocampo and Vos, 2008). Within this broader landscape, a gradual reconfiguration of innovation activities is taking place in which many developing countries are beginning to play a larger role.[1] For example, in 2006, China accounted for 36 per cent of all value added manufactured goods produced worldwide (UNCTAD, 2007), and since 1999 it has been steadily increasing its expenditure on R&D by an average of about 24 per cent per annum. As a result, its R&D/GDP ratio more than doubled in a decade, to reach 1.34 per cent in 2005 (UNCTAD, 2005). Similarly, India was ranked immediately after the United States as the second most favoured location for offshore R&D globally (Economist Intelligence Unit, 2007). However, a large number of other countries (especially, least developed countries) have not been able to keep pace.

These divergences have been attributed largely to the wide disparity among countries in technology and innovation capabilities, which is greatly impeding their ability to promote equitable growth and reduce poverty in a sustainable way. Uninterrupted and reliable energy supply is an important stimulant of innovative capacity and economic growth. As mentioned in chapter I of this Report, studies suggest a very high correlation between physical infrastructure (particularly energy) and industrial development in countries at low levels of development. A survey of a large number of developing-country firms in Africa concluded that the lack of reliable power supply is a crippling bottleneck in developing countries' industrialization efforts (Ramachandran, Gelb and Shah, 2009). This conclusion was confirmed by a recent study in Bangladesh that collected empirical data for the period 1973–2006, which found a direct causal relationship between the low supply of electricity and stunted economic growth (Sarker and Alam, 2010).

The lack of reliable power supply is a crippling bottleneck in developing countries' industrialization efforts.

At the same time, technology and innovation capabilities are important for promoting R&D and innovation to produce state-of-the-art RETs on the one hand, and for creating a critical base of knowledge (including the requisite physical infrastructure, technical maintenance and managerial capacities) required for adapting and disseminating RETs, on the other (UNFCCC, 2010). A critical threshold of technological capability is also a prerequisite for making technical improvements to RETs, which enable significant cost reductions so that they can be deployed on a larger scale in developing countries. The success of RETs-related technology transfer initiatives also depends on the ability of actors in developing countries to absorb and apply the technologies

Technology and innovation capabilities are important for promoting R&D and innovation to produce state-of-the-art RETs... and for adapting and disseminating RETs.

The absence of, or limited, technological and innovation capabilities is likely to constantly undermine national strategies for the greater use of RETs.

transferred. The absence of, or limited, technological and innovation capabilities is therefore likely to constantly undermine national strategies for sustainable development aiming to promote the greater use of RETs.

In 2006, developing countries as a whole accounted for 23 per cent of new investments in renewable energy sources worldwide (GTZ, 2007),[2] but a large share of these investments went to the more technologically advanced developing countries such as Brazil, China and India. Recent data for 2010 show that China recorded the largest new investments in renewables and related technologies (amounting to $49 billion), followed closely by Germany ($41.1 billion), the United States ($30 billion), Italy ($14 billion) and Brazil ($7 billion) (see table 2.1, chapter II). India ranked eighth with total new investments of $3.8 billion, and all of Africa accounted for just $3.6 billion (REN21, 2011). These figures underscore the variations in RET use and innovation among developing countries. Countries such as Brazil, China, India and South Africa are making significant technological advances in certain RE industries in the wind and solar sectors. However, progress in manufacturing, research, development and adaptation in most other developing countries and LDCs has been limited due to their relatively weak innovation capabilities.

Developing countries therefore find themselves at a crossroads on the issue of how to use established RETs, especially, wind, solar and biomass to successfully alleviate energy poverty, deploying them to complement traditional energy sources in the short and medium term, and even replacing those sources in the long term (UN/DESA, 2011). Issues of technological change and innovative capacity are at the forefront of this discourse. Countries' capacities for technological absorption need to be strengthened through coordinated policy support, but an additional priority should be to make existing technologies available and to assist in their greater diffusion. Innovation in RETs is moving at a fast pace globally, but there are some questions to which there are no clear answers at this point in time. As estimates presented in chapter II of this Report show, global projections of future energy supply vary widely. Left to its own, or left to the "market", it is unclear to what extent the surge in RETs will continue globally, and to what extent it will bring down the prices of these technologies for use at the individual household and firm level in the medium term. Public policy therefore has an important role to play in this regard, in addition to tipping the balance towards energy mixes that give prominence to RETs development in developing countries.

This chapter presents a framework for technology and innovation in developing countries, which is a necessary pre-condition for ensuring the greater use and innovation of RETs. It calls for: (a) a greater integration of RETs within socio-economic development strategies of countries; (b) the creation of capacity for increased technology absorption in general, and in RETs in particular; and finally (c) express policy support aimed at significantly integrating RETs into the national energy mix by tipping the balance in favour of RETs development, production and use. Section A presents an innovation systems perspective with regard to RETs. Section B describes the mutually dependent relationship between technology and innovation capabilities and the wider dissemination and use of RETs. It also shows the importance of promoting energy access as a necessary step towards economic development in developing countries. The chapter argues that greater policy intervention and support within countries aimed at all three areas listed above will help create the requisite economies of scale in use and diffusion that are required at the global level to drive down the prices of established RETs. At the same time, access to energy is also key to achieving many of the MDGs because of its impact on social development in terms of education, health and gender equality, among others (IEA, 2010). Section C suggests that there is also a need to integrate RETs use into policies and programmes aimed at poverty reduction and job creation, especially for the more economically vulnerable groups in developing countries and LDCs.[3]

B. TECHNOLOGY AND INNOVATION CAPABILITIES FOR RETs DEVELOPMENT: THE CONTEXT

Theory and evidence point to several results that now form the basis of international policy debates on innovation for development. First, innovation (as opposed to information or even knowledge) is the result of the interactions between firms and other organizations within a system, shaped by social, economic, political and historical factors (also called systems of innovation). Second, within countries and systems, accumulation of knowledge for technological change and innovation depends on a learning environment that promotes interactive learning. This is the process through which diverse actors in both the public and private domains communicate and collaborate for the creation, use and dissemination of new knowledge (Johnson and Lundvall, 2003). Accordingly, ways and means through which knowledge is perceived, applied and transformed are all influenced by socially and historically conditioned human perceptions and notions. This makes innovation – defined in the sense of "frontier" innovation and R&D, or as an incremental process leading to the building of technological capabilities – as much a social and cultural process as a scientific and technological one. Third, no technology, no matter how simple or complex, can be fully expressed in terms of its material value and components (Nelson, 1987). The unwritten, tacit (not easily embodied) knowledge explains why, when two producers in different parts of the world use the same technologies, there is always a discrete possibility that they may branch out into different applications, thereby producing completely different results. This focuses attention on a critical causal relationship between the availability of technologies and the importance of processes that underlie technological absorption capacity within countries.[4] Policy studies from developing and developed countries show that common constraints hinder knowledge accumulation (box 3.1).

There is a critical causal relationship between the availability of technologies and the processes that underlie technological absorption capacity within countries.

Box 3.1: Policy-relevant insights into technology and innovation

Studies in evolutionary economics and technical change stress the need to view technical change and knowledge accumulation within countries from a systemic perspective. The "national systems of innovation" framework, which derives from concepts of evolutionary economics, is based on the premise that technological change and knowledge accumulation that contribute significantly to economic development across countries are systemic in nature. Also known as the systems failure approach, it emerged in the works of Freeman (1987, 1988), Lundvall (1985 and 1992) and Nelson (1993). It originated primarily as a tool to help explain how country-specific factors contribute to innovative performance and economic growth,[a] and initially focused on the impact of improved coordination between R&D institutions and other secondary research units on the production system in its early stages (Lundvall, 1985).[b]

A basic premise of the concept of systems of innovation is that firms operate, grow and innovate within a network of firms and actors through interactive learning, which is essentially incremental and cumulative, leading in turn to the accumulation of new in-house capabilities. Both market and non-market mechanisms mediate interactive learning and innovation, and therefore market failure is only one component of the system. This failure can be rectified by altering the costs and benefits or payoffs of R&D to firms. As opposed to this, a system failure occurs when market and non-market agents interact in sub-optimal ways, or do not interact at all, and when critical actors within an innovation system do not promote innovation (see, for example, OECD, 1998).

Source: UNCTAD.

a The term "national system of innovation" was used first by Freeman (1987), when conducting an analysis of Japan's economic performance and growth. Nelson (1993) compares institutions and mechanisms that support technological innovation in 15 countries.

b The approach relies largely on evolutionary economics, although some scholars emphasize other theories, such as the theory of interactive learning and Schumpeterian economics.

There are several unique features of technology and innovation in RETs compared with other sectors on which many policy studies have focused. First, there is already a well-established energy system globally, and RETs are technologies that seek to provide alternative solutions to achieve the same result using natural resources of a different kind (such as sun, wind and water). Their unique selling point is that they offer environmentally friendly solutions to energy needs for the same service, namely the supply of energy. This is different from innovation in other sectors where competition is structured around the provision of newer products and services at reasonable prices.

Second, RETs demonstrate intermittency issues, as pointed out in chapter II, which calls for a systemic approach to promoting innovation in the sector. The systemic dimension is important to balance steady sources of energy that are largely predictable (such as availability of sun for a certain number of hours each day) with other sources of energy supply that may be more variable (such as wind intensity). Evidence shows that intermittency of different renewable energy supplies can be dealt with quite easily within electricity supply systems when solutions are designed from a systemic perspective.[5] A systemic treatment of renewable energy technologies is important from another perspective, namely the management of demand for energy, which is at least as important as the consideration of different sources of energy in the energy matrix of countries. Work regularly undertaken by organizations that promote RE-based solutions in developing countries (some of which were consulted for this *TIR*) shows that demand for energy in rural areas can be met very effectively by rural off-grid energy solutions using RETs (box 5.2 in chapter V).[6] The end-use dimension (i.e. how many people can access a particular supply and how effectively it can be provided) will need to play a major role in considerations of RETs as a means of alleviating energy poverty in developing countries. A systemic perspective, as proposed in this chapter, helps also to consider the demand dimensions when designing on-grid, off-grid or semi-grid applications using RETs, depending on the needs of countries.

Technology and innovation capacity is fundamental for making minor technical improvements that could enable significant cost reductions in production techniques, adaptation and use.

Third, it is often assumed, incorrectly, that technological capacity is required primarily for R&D aimed at the creation or development of newer RETs. As this chapter shows, technology and innovation capacity is fundamental for other aspects of RETs as well, such as:

(i) Making minor technical improvements that could enable significant cost reductions in production techniques, adaptation and use; and
(ii) Adaptation, dissemination, maintenance and use of existing RETs within key sectors of the economy, which depend not only on the availability of materials, but also on diverse forms of knowledge (IPCC, 2007).

Fourth, in developing countries, there is an urgent need to promote choices in innovation and industrial development based on RETs. These choices may differ among countries depending on their specific conditions and the kinds of renewable energy resource(s) available. The specific characteristics of different RETs, varied project sizes and the possibilities for off-grid and decentralized supply, imply many new players, both in project development (new and existing firms, households and communities) and in financing (existing lenders, new microcredit scheme, government initiatives).

An innovation systems perspective for RETs helps to map the various kinds of competencies that interact in the processes of R&D, innovation, adaptation and use of RETs. Linkages between the actor networks not only help to foster learning in various RETs, but are also essential for the integration of RETs into other sectors of the economy, such as transport and construction. Viewing RETs from an innovation perspective helps to establish the key linkages that need to be fostered through policy intervention. The networks and linkages that play this important role are explored in this section.

1. Key networks and interlinkages for RETs

a. *Public science through public research institutions and centres of excellence*

Public science (i.e. scientific research funded by governments) has been a major source of new knowledge, and has played a critical role in the emergence of several new technologies, including RETs, that are of increasing importance to the global economy. The positive relationship between academic research and innovation capabilities is well established, with the caveat that several factors condition the extent to which public science leads to technological innovation. Complementarity between public science and technological innovation varies among sectors: scientific research may be more applicable to technological advances and product development in some sectors than in others (see, for example, Andersson and Ejermo, 2004; and Dosi et al., 2006).

In the specific case of RETs, a large number of mature technologies currently in use and many others that are being developed (including electric cars) rely extensively on primary and applied research feedback loops with universities and educational centres of excellence (Stäglich, Lorkowski and Thewissen, 2011). In industrialized countries, whereas public funding for research in high technologies has been gradually shrinking, it is still the mainstay of research and innovation for RETs.[7] Historically, public science carried out in universities, research institutions and centres of excellence has aimed at expanding the science base for new and risky technologies that were not fully mature or that were applied in various sectors of production (box 3.2). Public-funded research is equally relevant for RETs for a variety of purposes, including for use and adaptation, incremental technological improvements and new scientific breakthroughs for application. For convenience, they can be classified into the following categories (ISPRE, 2009; see also Henzelmann and Grünenwald, 2011):

(i) *Energy efficiency*, including better conversion efficiency, performance, reliability and durability;

RETs rely extensively on primary and applied research feedback loops with universities and educational centres of excellence.

Box 3.2: The role of public science in building technological and innovative capacity

Public research institutions serve two main functions in the development and maintenance of competencies. First, they act as primary centres of innovation when various technical disciplines are initially introduced. They foster interactive learning between public and private institutions by: (i) promoting a product/market focus in innovation efforts in public sector institutions that often tend to be disconnected from product development; (ii) increasing mobility of skills and personnel between public and private sector institutions; and (iii) attracting funding from the private sector for important research programmes. When innovation systems are sufficiently developed, public science continues to perform a supportive role at two critical stages of the R&D cycle. In research, public science provides much needed state-of-the-art applied research support to smaller firms engaged in niche areas. It also provides substantial support services to universities with a specialized focus, matching advanced laboratory facilities and human resources, and in many instances it promotes interaction between universities and industry.

Whereas some areas of social science (such as human capital theories) have viewed schooling as a factor in enhancing the productivity of workers, recent studies on innovation have begun to unravel the extent to which schooling, higher education and industrial productivity are correlated (see, for instance, Gehl Sampath, 2010). It is still difficult to map the impact of certain kinds of educational investments made by countries as they proceed along their technological trajectories, because opportunities are driven by technological breakthroughs, markets, customer preferences, and, most importantly, by the ability of the innovation system in a country to respond rapidly to such stimuli. But clearly ex-ante decisions on various aspects of innovative capacity, such as a country's schooling system, its preferences for secondary and tertiary education (e.g. whether there should be a greater emphasis on natural sciences or other disciplines, or whether there should be centres of excellence for tertiary education) and investment in public sector research, all have an impact on the technological absorptive capacities of countries and sectors.

Source: UNCTAD.

(ii) *Material efficiency*, including advanced manufacturing techniques for components that substitute expensive with cheaper and reliable material inputs and reduce the use of toxic materials;
(iii) *Sustainable management*, including sustainable production processes that can reduce environmental impacts of manufacturing, use and final disposal;
(iv) *Storage efficiency*, including better methods for grid storage and integration of RETs into existing distribution systems;
(v) *Technological change and development*, including new mitigation and adaptation technologies (UN/DESA, 2009); and
(vi) *New R&D* into state-of-the-art RETs.

In many developing countries, an increasing share of public science activities is being directed towards strengthening capabilities in areas related to RETs.

In many developing countries, an increasing share of public science activities is being directed towards strengthening capabilities in material sciences, chemistry, engineering and physics in areas related to RETs. For example, in India, the Council for Scientific and Industrial Research (CSIR) accounted for more than 30 per cent of all green patent applications filed between 2000 and 2007 (OECD, 2010).

Given the knowledge-intensive and interdisciplinary nature of RETs research, universities are important for promoting innovation for the following reasons:

(i) They are able to provide the requisite level and quality of human skills (amount and quality of researchers), train new people and conduct research.
(ii) They possess the appropriate laboratory technology and equipment to train students in interdisciplinary areas of research.
(iii) Public research serves a coordinating function for promoting interaction between researchers engaged in the same/similar or related fields in universities and public institutions within the country and abroad.

Newer breakthroughs in public-funded research can help achieve cost reductions in RETs.

Other specific channels for public science exist in the case of RETs. For already proven technologies, such as wind and water-based RETs, the private sector is actively engaged in R&D that seeks to make them more adaptable to different situations. It can be supported through short-term, public-funded R&D in developing countries aimed at improving receptivity and acceptance of such technologies in the domestic context. In the case of some other RETs, such as solar PV installations, both basic and applied research are required to achieve cost-cutting in the technology, such as through higher conversion efficiency, lower consumption of materials and use of cheaper inputs in the manufacture of PV panels. Newer breakthroughs in public-funded research can help achieve cost reductions in such RETs, so as to ensure that those RETs can compete (in the absence of subsidies) with established fossil-fuel sources of energy "…once external environmental costs and other contributions to social goals (e.g. access, security) are taken into account" (ISPRE, 2009). University research also plays a key role in tackling socio-economic challenges that inevitably arise in the process of demonstrating and deploying new technologies. In the case of modern biomass technologies, there are still numerous environmental challenges that public science can help to resolve.

b. Private sector enterprises

Innovation, ideally induced by competition, creates winners and losers. Innovative firms have a negative externality on firms that do not innovate, and the latter therefore lag behind the state-of-the-art innovators in any field of technology. Such an externality gets internalized through the market mechanism, since innovation induces changes in production functions that result in lower costs of production, so that firms that do not upgrade fail to compete. The process of innovation based on constant competition promotes social welfare, and therefore the primary aim of policies should be to promote competitive environments for firms to thrive within sectors, RETs being no exception to this general rule.

The degree of vertical integration in sectors is often related to its technological characteristics.[8] R&D in RETs is becoming increasingly globalized[9] though a large share of all new product/process development is undertaken by the industrialized countries. In wind technology, 8 of the 10 leading manufacturers of wind turbines are European, including companies such as Vestas (Denmark), Enercon (Germany), Gamesa (Spain), GE Energy (Germany/United States), Siemens (Denmark/Germany), Nordex (Germany), Acciona (Spain) and Repower (Germany). Despite this, several new technological entry points have begun to open up for developing-country firms that are seeking to specialize in one or more aspects of RET production processes. The solar PV industry, for instance, is quite fragmented. A United States-based company, First Solar, is the leading firm worldwide, but firms from China, such as Suntech, Yingli Green Energy and Motech Solar, are rapidly expanding their market shares globally at the expense of already well-established German and Japanese firms (Hader et al., 2011).

There are also growing opportunities to participate in value chains for RETs, such as those for solar PV technologies that are fast emerging globally, in order to adapt and reduce costs, attain economies of scale of production and access untapped markets.

For wind energy, two slightly varied market trends are emerging. While mature markets are focusing on larger turbines and offshore installations, developing countries are focusing on smaller products and onshore installations. A second important trend is that the market for provision of wind energies by utilities is gradually expanding, thereby creating potential manufacturing possibilities along the value chain for small and medium-sized players, including independent power producers. New entrants are emerging, especially in China, the Republic of Korea, and some other Asian countries, including India, to tap into these technology entry points. These local suppliers are providing products along the value chain, and increased outsourcing is foreseeable, especially given the need to cut production costs further (Hader et al., 2011). Table 3.1 provides a list of emerging specialization and entry points for developing-country firms dealing in wind and solar technologies (see also box 3.3 for examples from China and India). This table lists some of the technological niches in which firms from developing countries have accumulated capabilities and emerged as suppliers globally.

The possibility to participate in global value chains, or other entry points, as discussed here, could provide a useful means for firms in developing countries to potentially accumulate new technological capabilities

New technological entry points have begun to open up for developing-country firms seeking to specialize in one or more aspects of RET production processes.

Table 3.1: Relative specializations and potential entry points for firms in wind and solar energies

Technology	Technological sophistication and entry points	Developing countries with significant capacities
Solar photovoltaic installations	Highly sophisticated, but with increased opportunity to specialize in niches along the value chain	Brazil, China and India
Multi-megawatt offshore turbines (wind)	Highly sophisticated	
Small turbines (wind)	Relatively sophisticated, with increased opportunity to specialize in niches along the value chain	China, India
Biofuels	Relatively sophisticated, especially for large-scale production	Brazil, China, India, Indonesia, Malaysia, the Philippines and Thailand
Biomass	Low sophistication, easy applicability	Bangladesh, China, India and Kenya
Low head turbines (hydropower)	Relatively sophisticated; potential opportunity for expansion exists but is currently limited by the low level of use in developing countries[a]	China[b]

Source: UNCTAD.

[a] The percentage of untapped hydropower globally is estimated at 65 per cent, whereas in Asia, Africa and South America, 90 per cent of total hydropower capacity is currently untapped (Hader et al., 2011).

[b] China is expected to become Asia's largest hydropower generator by 2015 (Hader et al., 2011).

Box 3.3: Examples of private firms in wind and solar energy: China and India

Solar: China is the world's biggest exporter of solar PV panels; around 95 per cent of its total production is exported to other parts of the world. In 2009, China exported over $10 billion worth of solar panels and cells, more than twice as much as the second biggest exporter and almost 80 times the value exported only 10 years earlier.[a] Suntech, the third largest solar company in the country had an annual production capacity of 1 GW in 2009. India also has several large solar manufacturers such as Moser Baer Photovoltaic Ltd, Tata BP Solar, Central Electronics Ltd and Reliance Industries. Indian firms manufactured solar PV modules and systems worth 335 megawatt power (MWp) up to March 2007, of which 225 MWp was reportedly exported. The Indian Government now plans to build the world's largest solar power plant in the state of Gujarat at an estimated cost of $10 billion, with an expected capacity of 3,000 MW.

Wind: China ranks second worldwide for installed wind capacities, with private firms using advanced technology for the production of wind turbines. Sinovel, a Chinese firm, is the third largest wind turbine manufacturer in the world, accounting for 3,495 MW of energy supply in 2009, and it is also China's largest wind turbine manufacturer. Three Chinese companies now rank among the top 10 in terms of market shares for wind power (Bouée, Liu and Xu, 2011), though they focus almost exclusively on meeting domestic demand. Goldwind, another large Chinese wind turbine company, has recently acquired a majority stake in Germany's Vensys in an effort to expand its know-how. India, currently ranked as the third largest wind producer worldwide, is following closely behind China. Indian companies supply many of the components required for the generation of wind energy worldwide. These components are mostly exported, and the Indian company, Sulzon, is the world's third largest supplier of components to wind power operators, with a 6.4 per cent share of the global market (BTM Consult, 2009). Sulzon operates in three continents to produce components for the entire supply chain. It has R&D facilities in Belgium, Denmark, Germany and the Netherlands.

Source: UNCTAD, based on Bouée, Liu and Xu (2011) for China and Kalmbach (2011) for India.

[a] UN Comtrade database (HS 854140: Photosensitive semiconductor devices, including photovoltaic cells whether or not assembled in modules or made up into panels; light emitting diodes).

Production and manufacturing possibilities need to be steadily augmented by means of a policy environment that promotes the accumulation of knowledge and capacity-building.

in order to gradually move up the innovation chain. Although these developments are positive, the results will not accrue automatically. Production and manufacturing possibilities need to be steadily augmented by means of a policy environment that promotes the accumulation of knowledge and capacity-building in order for firms to upgrade and progress technologically. Failing this, there is always a risk that a large number of firms in developing countries will be entrenched at the lower ends of global manufacturing chains, as experienced in several other sectors such as readymade garments and electronics. An enabling environment for innovation and technological upgrading is discussed in chapter V of this Report.

c. End-users (households, communities and commercial enterprises)

Households and communities could play an important role in both on-grid and off-grid installations of RETs. Modern biomass and off-grid installations of RETs are aimed at rural communities as the primary target group. Providing the means to cook, electricity for basic household chores and energy for men and women to engage in economic activities could help boost economic growth and development (see chapter V for examples). RETs could also promote newer sources of employment and greater prosperity in rural areas, even through small off-grid installations, such as for selling milk-based products or for storing important temperature-sensitive drugs, as well as for ICT-based applications. For some other RETs, such as solar PV installations or utilities based on wind power, households are important actors. In countries, such as India and Tunisia, use of solar PV panels by individual households is on the rise. Energy providers also target households with new energy-mix schemes that combine intermittent supplies, such as wind power, with conventional sources. Commercial buildings, especially business and office spaces, account for a large amount of energy usage and could be very important user communities for RETs in developing countries in the medium and long term.

2. Linkages between RETs and other sectors of the economy

Ultimately, developing innovation capabilities depends on the ability of agents to collaborate and cross-fertilize ideas and results across a broad range of disciplines and skills in firms and other organizations. Collaborative networks also ensure that knowledge is constantly accumulated and used through a combination of tacit skills and codified information produced in innovation processes and exchanged between public and private sector institutions through a dynamic, self-reinforcing process of capabilities formation.

The reasons for intensified networking vary. Determining factors include access to new forms of knowledge, shared risks as a result of escalating costs of innovation, and the leveraging of market and skills opportunities. Inter-firm and inter-organizational flows of knowledge and skills in a user-producer relationship could take various forms, including the movement of skilled staff from one firm to another, subcontracting (manufacturing), licensing and joint ventures, franchises and collaborative agreements for marketing of products, and supplier-customer relations. Most importantly, asset pooling, be it in the form of human resources, finance or machines, is an important reason for collaboration.

Interactive learning in general depends on better linkages between university departments, centres of excellence and public research institutions, and those involved in product development, including the private sector. Other forms of knowledge interactions, such as those between foreign firms and universities, and between consumers, investors, developers and intermediary organizations – especially those that help gauge local demand, such as market research organizations – are also important.

In the case of RETs, establishing these interlinkages is important from several perspectives. Following from the discussion in chapter II, depending on the scale of the RET in question, different technologies will have different user profiles and markets within developing countries. These need to be carefully established and the linkages appropriately fostered to ensure that the adaptation and use of RETs deliver the expected benefits. Apart from complementing electricity generation, newer uses of RETs in different sectors of the economy are emerging. These are mostly associated with the drive to promote "green innovation", which denotes innovation conducted in an environmentally sustainable way. For instance, RETs are becoming more important in the transport sector, in building and construction, in battery technologies and in the chemical industry.[10] It is estimated, for example, that residences and commercial buildings in the United States account for 40 per cent of the country's total energy use (Reers, Benecchi and Koper, 2011). Similarly, in the automobile sector, it is increasingly clear that simply enhancing internal combustion efficiency in vehicles will not be sufficient to reduce carbon emissions (Staeglich, Lorkowski and Thewissen, 2011). Recent trends towards promoting the use of electric vehicles have forced a rethinking about the entire automotive industry value chain, including R&D in particular niches such as batteries, vehicle assembly, infrastructure, and new business models that guarantee care and maintenance (Henzelmann and Gruenenwald, 2011). This also implies greater possibilities for firms in developing countries to anticipate newer technological entry points related to RETs that are not necessarily limited to energy supply systems, as discussed in the previous section. These developments underscore the many positive externalities that RETs production and use can have for developing economies, depending on how these interlinkages are structured and fostered by countries. There are a few instances where RETs may have linkages with sectors of the economy that are not always positive, and where they compete with other needs such as in the case of biofuels (discussed in chapter II). Thus the costs and benefits of such linkages need to be balanced within national policy frameworks.

Apart from complementing electricity generation, newer uses of RETs in different sectors of the economy are emerging.

These developments underscore the many positive externalities that RETs production and use can have for developing economies.

C. PROMOTING A VIRTUOUS INTEGRATION OF RETs AND STI CAPACITY

There is urgent need for government action to change current patterns of energy use with reliable, established RETs.

Despite the various potential advantages cited with regard to the use of RETs, established fossil-fuel sources still dominate energy supply at present, providing up to 89 per cent of all global energy (Chichilnisky, 2009). A large proportion of the global population cannot afford these conventional energy supplies, as noted in chapter I of this Report, which makes the eradication of energy poverty an immediate goal for economic development.

According to estimates of the International Energy Agency (IEA, 2011), over 20 per cent of the global population (1.4 billion people approximately), most of whom live in rural areas, had no access to electricity in 2010. South Asia has the largest proportion of people without access to electricity (42 per cent of the world total), in spite of recent fast progress. Taking the entire population of this subregion, 38 per cent lack access to electricity, and 49 per cent of people living in rural areas lack access. In relative terms, sub-Saharan Africa is the most underserved region, with 69.5 per cent of the population having no access to electricity, and only a meagre 14 per cent of the rural population having access (table 3.2).

A large number of people in developing countries and LDCs (especially South Asia and sub-Saharan Africa) who lack access to affordable conventional energy sources rely on biomass (including wood, crop waste and charcoal), which continues to provide at least one third of all primary energy supply in these countries.[11] Use of such an alternative energy source is generally neither efficient nor healthy for the users and the environment. Therefore there is urgent need for government action to change current patterns of energy use with reliable, established RETs. While off-grid RETs (especially modern biomass-based) may be easier to deploy, others still remain very expensive at the scales required to make an impact in developing countries, despite rapid technological advances (UN/DESA, 2009). For example, a study by the IEA (2009) came to the conclusion that in the United States, electricity from new nuclear power plants was 15–30 per cent more expensive than from coal-fired plants, and the cost of offshore wind power was more than double that of coal, while solar power cost five times as much. Changing from the current global situation of no energy, or unreliable

Table 3.2: Access to electricity and urban and rural electrification rates, by region, 2009

Region	Number of people without electricity (millions)	Electrification rate (%)	Urban electrification rate (%)	Rural electrification rate (%)
Africa	**587**	**41.9**	**68.9**	**25.0**
North Africa	2	99.0	99.6	98.4
Sub-Saharan Africa	585	30.5	59.9	14.3
Developing Asia	**799**	**78.1**	**93.9**	**68.8**
China and East Asia	186	90.8	96.4	86.5
South Asia	612	62.2	89.1	51.2
Latin America	31	93.4	98.8	74.0
Middle East	22	89.5	98.6	72.2
Developing countries	**1 438**	**73.0**	**90.7**	**60.2**
OECD and transition economies	**3**	**99.8**	**100.0**	**99.5**
World total	**1 441**	**78.9**	**93.6**	**65.1**

Source: Reproduced from IEA (2010).

and often undesirable sources of alternative energy (such as traditional biomass), to one where industrial development adopts a cleaner growth trajectory is also essential for driving down the costs of RETs.

Mobilizing additional domestic resources in support of RETs will require the conscious development of policy strategies by governments all over the world, including overcoming different kinds of systemic failures inherent in the use of RETs. States, in designing institutional incentives, will need to play a fundamental role in tipping the balance towards energy sources that use RETs. Such incentives need to be designed and articulated at the national and regional levels so that collective actions can be fostered. Most importantly, energy production should cater to local needs and demand in countries, for which a systemic perspective is necessary. The International Renewable Energy Agency (IRENA) estimates that 40 per cent of all energy produced in Africa is exported, despite large-scale energy poverty in that region (see box 1.2, chapter I).

Government action will need to focus on two very important areas of intervention: addressing systemic failures in RETs, and tipping the balance away from a focus on conventional energy sources and towards RETs. Systemic failures in the RETs sector are varied and emerge from sources other than just the market; they can be caused by technological uncertainty, environmental failures or other systemic factors. Therefore, government intervention will be very important for addressing those failures. Similarly, while it is clear that there is a growing role for RETs as energy providers globally, government action will be critical for inducing a shift towards a wider application of RETs in the energy mix of countries.

1. Addressing systemic failures in RETs

The risks associated with the potential, viability and scale of application of RETs is due to four uncertainties: market-related, technological, general systems-related and environment-related. Both technological and market-related uncertainties tend to dictate firm-level actions and decisions for the building of capabilities in particular ways, which explains the varied performance of firms across sectors over time.[12] In the case of RETs, two other kinds of failure exist: systems-related and environmental, which also need to be taken into account.

Government action will need to focus on two very important areas of intervention: addressing systemic failures in RETs, and tipping the balance towards RETs.

Innovation across all sectors and industries requires investment, the returns on which are uncertain. Since innovation denotes the application of R&D results to create commercially viable products, demand plays an important role in returns on investment. Economic theory suggests that market failures caused by uncertain returns on investment can be corrected through a range of market-based instruments, including patents, tax incentives and subsidies. Government intervention in the form of industrial policy could minimize information asymmetries between user-producer networks, mitigate inefficient resource use and also address public goods issues.

Markets for RETs are only just developing, and forecasts of total market demand and market size vary depending on the assumptions made, not only with regard to the expansion of RETs per se but also to carbon pricing[13] and the availability of alternative sources of conventional energy, especially gas.[14] In such an environment, firms and organizations are faced with the choice of whether to invest in RETs as opposed to other technological sectors where returns are more secure (from a current perspective). Further technological uncertainty is caused by the constant flow of newer technologies that not only affect products and innovation cycles, but also consumer behaviour. Moreover, this also leads to a continuous reallocation of the technology-based strategic advantages of firms. In addition, changes in firms' organizational arrangements affect their technological opportunities and outcomes (e.g. Robertson and Langlois, 1995; Brusoni and Prencipe, 2001). Firms constantly need to compete and reorganize their internal strengths so that they are well prepared to exploit new technological opportunities presented by RETs.

Systemic failures exist as well, which undermine possibilities of expanding into RETs in developing countries. Most importantly, countries and sectors are path-dependent, and RETs face systemic risks of not being adapted, used or applied in other sectors of the economy. Manufacturing firms in developing countries are under considerable pressure in today's global trade environment to retain their competitiveness and export orientation. Therefore, policies that dictate a shift from conventional energy supplies to a mix of conventional and RETs, or purely RETs, to sustain their production will involve sunk costs. In the absence of political will and government and market-based incentives for firms to help offset such costs, such a shift will be difficult, especially for developing countries.

Combining conventional sources of energy with RETs is a policy choice that requires the mobilization of greater domestic resources.

Lastly, positive effects on the environment created by the use of RETs are not quantifiable. Besides, no single user/firm/investor has the incentive to take the risk to promote the use of RETs for the greater social good.

2. Tipping the balance in favour of RETs

Combining conventional sources of energy with RETs is a policy choice that requires the mobilization of greater domestic resources for innovation and technical change on the one hand, and sustainable pathways of development on the other. Both the policy framework and State intervention will play a decisive role in determining the future role of RETs and the appropriate mix of RETs and conventional technologies within a country.

Currently, as chapter II shows, RETs can sometimes be more expensive than conventional sources of energy, mainly because price estimates of conventional energies do not usually include the costs of grid connections and storage (which can considerably increase total costs). They also fail to reflect the environmental costs of these energies. Despite this, as noted in chapter II, average annual growth rates of capacity in the period 2005–2009 were between 10 per cent and 60 per cent for many RETs (IPCC, 2011). Globally, solar PV has grown the fastest of all RETs (over 60 per cent annually), followed by biodiesel production (51 per cent), wind power (27 per cent), solar water heating (19 per cent) and ethanol production (20 per cent) (Hader et al., 2011). Projections indicate that installed wind capacity will grow annually by 13 per cent worldwide until 2014, with a total installed wind capacity reaching an estimated 600 GW in 2020 (Hader et al., 2011).

In 2010, the amount invested globally in RET innovations equalled that spent on innovation in fossil fuel energy supplies, and it was greater than investments in nuclear energy innovations (table 3.3). This indicates increasing investments into RET innovations. However, much less is being invested globally in the diffusion of renewable energy when compared with the diffusion of other energy alternatives, and therefore this requires more emphasis.

Each time investment is made in generating more energy through RETs, not only does this result in a gradual shift in the energy

Table 3.3: Annual investments in global innovation in various energy sources, 2010 ($ billion)

Energy category	Innovation (research, development and deployment)	Diffusion
End-use and efficiency	>>8	300–3 500
Fossil fuel supply	>12	200–550
Nuclear	>10	3-8
Renewable energy	>12	>20
Electricity (generation and R&D)	>>1	450–520
Other, unspecified	>>4	1 000–5 000

Source: UNCTAD, adapted from Davis (2011).

base, but it also has a significant impact on the capacity of RETs to supply energy economically. For example, according to recent reports, every time the amount of wind generation capacity doubles, the price of electricity produced by wind turbines falls by 9–17 per cent (Krohn, Morthorst and Awerbuch, 2009; and UN/DESA, 2009). This holds true for all RETs: with each new installation, there is learning attached as to how the technology can be made available more effectively and efficiently in different contexts so as to lower costs over a period of time. According to UN/DESA (2009: 10), "...the more we learn about how to produce renewable energy, the less expensive it becomes". This effect has been demonstrated with regard to RETs over the past few decades: significant cost reductions have been observed with technological advances and growing usage.

The future expansion of RETs and their price competitiveness will depend on how and to what extent governments will proactively promote an agenda that combines (a) enforcement of carbon emission standards to reduce reliance on carbon-intensive technologies; (b) the use of RETs at domestic and industrial levels to complement existing sources of energy so that established technologies, such as solar PV, can rely on the economies of scale required to reduce the costs of production; and (c) improvements in the general technology and innovation capacities of countries to foster a virtuous cycle of RETs integration. Such a "big push strategy" or "tipping point" is important to lower the price of RETs, which will not fall rapidly on its own. It is also important to ensure that expanded markets for RETs result in greater investments in technological improvements and increased production in order to achieve cost competitiveness (UN/DESA, 2009).

A range of policy opportunities exist to create synergies between R&D, technical change, production and dissemination for entrepreneurs as well as end-users in developing countries, as discussed in chapter V of this *TIR*. The role of governments is critical for making RETs feasible at each of these entry points. Government agencies and the policy framework can play a decisive role in the following ways:

(i) Promoting the general innovation environment for the development of science, technology and innovation;
(ii) Making RETs viable; and
(iii) Enabling enterprise development in and through RETs.

a. Government agencies and the general policy environment

Total grid-based electricity capacity using RETs was estimated to amount to 3,400 GW in 2000, of which 1,500 GW was attributable to developing countries (Martinot et al., 2002).[15] This capacity has been expanding steadily, and developing countries have been investing in different kinds of RETs based on relative endowments (see the case of wind power in Chile, discussed in chapter V). Further integration of RETs into national development strategies of countries needs to be supported by express policy incentives that promote learning-by-doing, learning-by-using and networking opportunities, all of which affect the cost and value of RETs (Jaffe, Lerner and Stern, 2005; Skoglund et al., 2010). Government support and the general policy environment are important for fostering STI capacity, given the mutually dependent relationship between RETs and the STI environment. The general policy framework needs to address a range of constraints on innovation in developing countries (box 3.4).

Supportive policy frameworks that remove, or at least help to overcome, some of these constraints on technological change are important for several reasons. Universities and public research can perform a range of short- and medium-term support functions, as identified in the previous section, for example by providing ways and means to adapt existing RETs instead of, or in conjunction with, the private sector. They can also create awareness of RETs and promote their acceptance by people in developing countries. The presence and

Each time investment is made in generating more energy through RETs... it has a significant impact on the capacity of RETs to supply energy economically.

Future expansion of RETs will depend on how governments will proactively promote an agenda that combines the enforcement of carbon emission standards alongwith the wider use of RETs.

Box 3.4: Constraints on technology and innovation in developing countries

The main constraints on technology and innovation capacities in developing countries can be categorized as follows:

(a) *Lack of local capacity to absorb and use knowledge*, primarily determined by the availability of human skills locally and the institutional capacity of the system to provide the basis for innovative activity within any of the four knowledge domains identified in the previous section. In the absence of this, access to knowledge remains at best just access to information, since the actors lack the capacity to build further upon it.

(b) *Lack of well-developed institutional frameworks to forge second-best responses to innovation issues*, which manifests in the form of high transaction costs to conduct innovation activities. Institutional frameworks that are either incomplete or do not clearly specify the roles and responsibilities of various actors often result in organizations being set up with overlapping competencies and duplication of, or gaps in, roles and responsibilities.

(c) *Lack of resources in the general innovation environment*, which includes lack of physical and knowledge infrastructure, as well as financial instruments for reducing innovation risks. Innovation processes are associated with their own range of technological and market-related uncertainties, but at the same time innovation outcomes can vary when the same activities are conducted by diverse groups of individuals in different contexts whose levels of "imagination and accuracy" differ. This largely explains the varying performances of firms and sectors (Archibugi and Michie, 1997). In resource-constrained developing countries, there are few, if any, institutions that reduce market- related uncertainties and promote innovation.

(d) *Lack of a supportive public sector* that has the human and financial capacity to conduct relevant basic and applied research and industrial R&D. This constraint can have very different consequences for different sectors. In sectors that require the involvement of publicly funded research, such as pharmaceuticals, agriculture and new technologies, an efficient and well-endowed public sector is a prerequisite for innovation.

(e) *Lack of a thriving private sector* that can uptake results of industrial R&D conducted in public sector organizations is a common constraint on innovation in developing countries.

(f) *Lack of collaborative linkages* that allow mobility of ideas and human capital between firms and organizations alike. Competing agendas of organizations involved in STI, lack of a collaborative culture amongst academics and industry practitioners, lack of incentives that reward collaborative conduct, and lack of discernable benefits of collaborative linkages within the system, all contribute to poor or no collaborations, and therefore to the absence of interactive learning.

(g) *Lack of policy competence* in developing countries is perhaps as complex a phenomenon as the lack of innovation capability itself. Governments, by their actions as well as inactions, make technology choices for national development. They should be able to identify market failures and opportunities, make strategic choices, translate them into policies and ensure effective implementation of those policies.

Source: UNCTAD, based on Gehl Sampath (2010).

Deployment-related learning fosters skills that are essential to maximizing the efficiency of use of RETs.

availability of skilled human resources and a conducive innovation environment are important for promoting what is increasingly known as deployment-related learning in RETs. Deployment-related learning fosters skills that are essential to maximizing their efficiency of use. The final cost of RET use, which will be decisive from the consumer's perspective, depends not only on the cost at which the RET is available for the first purchase (for example, a solar panel) but also on the cost of maintaining and sustaining it over time. In the case of solar PV installations, for example, a cost breakdown shows that only 50 per cent of the total cost is for the PV cells, and the remaining costs are split between installation costs (30 per cent) and maintenance (20 per cent). Deployment-related learning is also important for finding new and locally suitable means to connect RET-based energy supplies with grid or mini-grid applications as well as for ensuring energy efficiency of use and storage. The installation and use of solar PV requires skills in roofing and electrical engineering, which are needed to ensure that the energy source is fully connected to a grid for usage. In the absence of reliable maintenance, loss of energy is common and the costs of shifting to RETs increases.

b. Facilitation of technology acquisition in the public and private sector

An increased absorptive capacity of the innovation system as a whole, involving local actors in the public and private sector, is a prerequisite for local adaptation and use of existing technologies in the renewable energy sector, as well as for other innovative pursuits. In addition, the policy framework needs to proactively support the acquisition of RETs by establishing a legal and institutional environment that promotes the expansion of RETs-based private sector activity. All policy efforts aimed at technology transfer and technology sharing should actively seek to engage the private sector. There are several impediments to developing-country firms accessing existing technologies. Searching for technology suppliers can be a costly and time-consuming process. Negotiating for the rights to use certain technologies can also require skills in legal and managerial capacity, which may not be easily available to firms in developing countries. These competencies need to be fostered through the general innovation policy framework. Furthermore, as chapter IV shows, there seems to be an increasing trend toward IPRs protection of climate-friendly technologies. It is not clear whether and to what extent such IPRs will affect the acquisition of technologies by firms and private sector organizations in developing countries in the future. Therefore it may be important to design national IPR regimes in ways that they do not impede technology acquisition priorities of countries (see chapter V).

c. Promotion of specific renewable energy programmes and policies

Several of the larger developing countries have initiated large-scale renewable energy programmes in an effort to harness the potential of alternative sources of energy. Apart from China and India, both of which targeted renewable energy use to supply at least 10 per cent of their total demand by 2010, other countries, including Brazil, Croatia, Egypt, Jordan, Mexico, Morocco, the Philippines, South Africa, Thailand and Tunisia, and have RETs programmes (Siegel, 2006).

Newer projects, some of which are regional in nature, are in the process of being implemented in many other countries, and they could contribute significantly to alleviating energy poverty in those countries. A promising venture is the Turkana wind corridor project currently under way in East Africa (see chapter V).

d. Attainment of grid parity and subsidy issues

Several of the larger developing countries have initiated large-scale renewable energy programmes in an effort to harness the potential of alternative sources of energy.

For grid-based usage, the feasibility of RETs adaptation and use on a wide scale depends on the attainment of grid parity. This is the level at which the renewable energy source is equal to or cheaper than the other conventional sources of electricity. Such grid parity depends on the RET in question, and is determined not only by its technological characteristics, but also by regional differences in cost and performance, infrastructure limitations and discount rates (IPCC, 2011, figure 5)[16]. Experiences of several industrialized countries show that government incentives and subsidies often play a very important role in the achievement of grid parity. A case in point is solar PV energy, which owes its development to the efforts of the governments of Germany and Japan, both of which began to massively subsidize PV technologies in the 1990s.[17] Other countries followed, leading to a wider, much broader application of solar-based RETs.

Experiences of several industrialized countries show that government incentives and subsidies often play a very important role in the achievement of grid parity.

Governments have also subsidized the adoption of RETs to a very large extent. Evidence available on the dissemination and use of RETs in countries shows that the differences in scale of use of solar technologies in India or the Republic of Korea, or even France, compared with what is observable in Germany and Spain is largely due to the amount of subsidies and user incentives offered by the governments of the latter countries. According to the analysis in chapter II of this Report and that of the IPCC (2011), some RETs are clearly emerging as competitive alternatives to energy in several markets.

e. Promoting greater investment and financing options

Different kinds of RETs require different scales of ex-ante investments, and technological characteristics dictate the kinds of support infrastructure required.

Different kinds of RETs require different scales of ex-ante investments, and technological characteristics dictate the kinds of support infrastructure required. It is projected that European countries, for instance, will invest €120 billion in wind energy development alone by 2020 (Hader et al., 2011). While it is unrealistic to assume that developing countries and LDCs will be able to make similar investments in RETs, technological choices often rest on several known and unknown parameters. Amongst the known ones, both wind and solar energies are subject to fluctuations, and without "enablers" they are unable to provide a steady, reliable source of energy. Since solar energy cannot be generated at night and wind speeds are unreliable, smart grids are important infrastructural requirements to store energy for use on a larger scale. While some technologies for storage already exist, others are being developed, and future developments may reduce the costs of some of the options currently available.[18] Policy choices will need to be made that will mix intermittent energies generated by RETs with other steady sources of energy, which could be RET-based or conventional energy-based (Delucchi and Jacobson, 2011). At the same time, there are many unknown factors. For example, what could be the potential cost reductions of RETs in the medium term? Will solar thermal or solar PV take the lead in world markets? The answers to these highly relevant questions are currently only guesstimates, despite the trends in cost reductions of RETs. Therefore, it would be useful for developing countries to consider promoting a basket of RETs, as suggested in chapter II. In particular, they need to explore greater financing options for bolstering enterprises' small, medium and large-scale RETs projects. Judging by current trends, it seems that international financing for climate change mitigation efforts focuses on large-scale projects (see chapter IV) and increasingly on small-scale initiatives (UN/DESA, 2011; Hamilton, 2011). However, there is also a need to provide financial support for the expansion of medium-sized projects by the private sector in developing countries.

Policy choices will need to be made that will mix intermittent energies generated by RETs with other steady sources of energy, which could be RET-based or conventional energy-based.

Facilitation of foreign direct investment in the sector would also be important, and, if designed and implemented well, it could result in greater investment and technology transfer to host countries. This involves creating so-called enabling conditions to make the host country an attractive environment for investors as a key goal of the broader innovation policy framework for RETs. Investors, both foreign and domestic, consider a number of factors when making decisions. They assess the risks and difficulties of investing in a given country in products using any given technology, and add this to the expected costs. Broadly, the factors investors consider include political and macroeconomic stability, an educated workforce, adequate infrastructure (transportation, communications, and energy), a functioning bureaucracy, rule of law, a strong financial sector, as well as ready markets for their products and services. Many factors contribute to shaping a country's national energy policy – including its policy on RETs – such as history, politics, geography (natural resource endowments) and chance, on the one hand, and innovation and production climate on the other. Studies have noted that many developing countries, particularly the least developed among them, are not getting their full share of investments for the development of renewable energy because existing national policies do not render such investments attractive for most projects (Amin, 2000; Chandler and Gwin, 2008; Point Carbon, 2007; Dayo 2008; Neuhoff 2008; Cosbey et al., 2009).

f. Monetizing the costs of energy storage and supply

Monetizing the costs of energy storage and supply of conventional energy sources, along with an estimate of the environmental costs of using such energy, will make it easier for RETs to compete. Countries and consumers will base their ultimate choice of energy sources not just on the costs of supply alone, but rather on a combination of price competitiveness and the costs of integrating the new sources into current modes of operation, along with environmental and social considerations (IPCC, 2011).

3. Job creation and poverty reduction through RETs

The current state of underdeveloped energy infrastructure in developing countries could be partially remedied through the use of RETs. Not only could RETs potentially help reduce energy poverty; they could also reduce social inequalities through the creation of new jobs in their application. Germany, for example, created 40,000 new jobs in the RE sector (particularly for electricity) between 1990 and 2002, and these are projected to increase to 250,000–350,000 by 2050 (Holm, 2005). Some of the main barriers to greater market penetration of RETs in developing countries are the lack of trained installers/service craftsmen and an absence of national standards/testing facilities, all of which have the potential to create jobs if training opportunities are provided. For example, it has been estimated that if South Africa were to use RETs in generating just 15 per cent of its total electricity by 2020, a total of 36,400 new jobs could be created without reducing employment in the coal-based electricity sectors (AGAMA Energy, 2003). It is estimated that other RETs, such as solar water heating and sustainable biomass production, have greater potential for direct job creation, the latter being particularly labour-intensive (Holm, 2005). More generally, simply the provision of greater access to energy through RETs would help to improve the income-generating capacity of the poor in three important ways: by creating new income-generating opportunities (such as working on electrical machinery), improving efficiency and productivity of existing opportunities (such as replacing manual tailoring with electric tailoring machines) making them more profitable, and, finally, reducing the time spent on existing chores such as women's daily collection of fuelwood for cooking (Practical Action, 2010).

Furthermore, RETs can help promote the MDGs in various ways (Practical Action, 2010) including:

(i) By providing greater access to health care. For example health centres in remote villages in the Philippines, have solar powered refrigerators for storing medicines and vaccines for people from the neighbouring villages. This initiative, the result of a joint community-based project of the Australian and Philippine governments, requires residents to maintain the solar batteries on their own.

(ii) By providing greater access to ICTs. In several countries (Kenya being a good example), access to energy through RETs has enabled greater penetration of ICTs into rural areas. The electricity required to power appliances and charge batteries for ICT use is made available through extension of the grid or through decentralized energy systems such as solar panels.

(iii) By promoting gender parity. Greater access to energy enables a large percentage of women in developing countries to reduce the time spent on household chores and to take on additional income-generating activities, which promotes gender parity. Women also find more time to participate in social and community activities, including improving their literacy rates.

Reducing inequality and poverty through RETs also requires rural enterprise development and small-scale financing, neither of which receives particular or adequate attention in discussions concerning RETs. However, there are some examples of successful support in these areas, such as that provided by the Grameen Bank in Bangladesh, and consumer credit for home solar systems provided by Grameen Shakti (Bangladesh), Viet Nam's Women's Associa-

Not only could RETs potentially help reduce energy poverty; they could also reduce social inequalities through the creation of new jobs in their application.

RETs can help promote the MDGs in various ways.

tion (Viet Nam), Sarvodaya (Sri Lanka) and Agricultural Financial Corporation (Zimbabwe). A survey of experiences in promoting RETs in the rural context shows that they are the most beneficial in contexts where economic development is already taking place (Martinot et al., 2002). Two lessons stand out. The first one is that local knowledge matters, as noted also by the Cancun Adaptation Framework of 2010. The Framework stresses that adaptation needs to be based on a combination of the best available science and the local and indigenous knowledge of communities (UNFCCC, 2010). Second, RETs are most easily disseminated when bundled with existing products, as this helps to lower costs for private users and small-scale industries that can arise from an abrupt transition to RETs. For example, dealers of farm machinery, fertilizers, generators, batteries, electronics and electrical utilities can all bundle their services with RETs in order to make their offers more easily acceptable to consumers. But poverty eradication is not a direct, automatic consequence of RETs, as is often assumed; it requires clear, express policy action by governments that link the use of RETs to poverty reduction as much as to the reduction of energy poverty.

Eliminating poverty through RETs requires clear policy action by governments that promotes poverty reduction hand-in-hand with universal energy access.

D. SUMMARY

This chapter has contextualized the technology and innovation issues relating to promoting the generation, adaptation and use of RETs. The analysis shows that developing countries and LDCs may have different needs and capabilities in using existing technology and innovation capacity to support the expansion of RETs in their economies. Despite this, several common issues remain, which are applicable to all developing countries. The successful use of RETs will depend on the ability to integrate them into countries' innovation strategies in order to reap maximum synergies for sustainable development. This calls for governments to adopt an agenda of proactively promoting access to energy services of the kind that is conducive to development, while also focusing on the important positive relationship between technology and innovation capacity and increased use of RETs. The chapter also shows that technology and innovative capacity are critical not only for RETs production and R&D-based innovation, but also for adaptation and greater use of RETs. Regular maintenance, adaptation and incremental improvements to RETs suited to local contexts could lead to their greater acceptance in developing countries, but this depends on local actors possessing some level of innovative capabilities. Innovation systems in developing countries are fundamental to shaping the needed capacity for the technological learning that is important for adaptation, use, production, R&D and innovation of RETs. Finally, the chapter stresses the need for greater mobilization of financial resources, in addition to increased access to the most advanced, cost-cutting technological improvements to established RETs. Greater international support for developing countries will be critical on both these fronts. At the same time, national policy frameworks should aim at harnessing the virtuous relationship between technology and innovation capacity of RETs for inclusive economic development, job creation, reduction of energy poverty and climate change mitigation. International policy challenges and sources of support are discussed in the next chapter.

NOTES

1 For a discussion on the reorganization of global innovation and the emergence of developing-country capacities, see Castellaci and Archibugi (2008) and Chesbrough (2003).

2 In 2006, total new global investments in renewable energy sources amounted to about $71 billion, an increase of 43 per cent over 2005. However, only $15 billion of this was invested in developing and emerging countries (GTZ, 2007).

3 The importance of including poverty reduction in discussions on the green economy and RETs is becoming increasingly clear. For example, UNEP (2011: 2) defines the green economy as an economy "[t]hat results in improved human well-being and social equity, while significantly reducing environmental risks and ecological scarcities."

4 With regard to innovation, scholars have long identified the relevance of country-level absorptive capacity. A firm's absorptive capacity lies in its ability to identify important sources of knowledge and technological change, route them into its internal learning processes and utilize them to build its own competitive advantages as an ongoing process (Cohen and Levinthal, 1990). Innovation systems of countries where firms are located are dynamic and dictate the process through which capabilities are formed.

5 It is estimated that electricity systems can easily handle up to 20 per cent of renewable energy, and even more if systems are designed with some adjustments in intermittency.

6 Based on consultations with The Energy and Resources Institute (TERI), India.

7 Examples include the Frauenhofer institutes in Germany and the Commonwealth Scientific and Industrial Organisation (CSIRO), a national science and research agency in Australia (Henzelmann and Grünenwald, 2011).

8 There is still substantial ambiguity in the literature as to whether greater technological change in a sector induces firms to increase or reduce vertical integration and how this can be studied (see Ciarli et al, 2008; and Dosi et al, 2006).

9 UNCTAD estimates that over 80 per cent of global R&D is conducted in just 10 countries, and most of it (including for technologies required for climate change mitigation) is directly undertaken by transnational corporations (UNCTAD, 2010).

10 A case in point is Dupont, which has entered the renewables business by creating renewable polyester (Dupont Sorona) and renewably sourced theroplastic elastomer (Dupont Hytrel RS).

11 Some estimates suggest biomass accounts for up to 45 per cent of all primary energy supply (e.g. Martinot, 2003).

12 See, for example, Archibugi (2001); Malerba (2002) and (2004); Marsili and Verspagen (2002); Dosi et al (2006).

13 This is discussed in detail in the next chapter.

14 For example, forecasts about whether or not the prices of solar PVs will fall over the next two decades vary according to whether the assumptions take into account greater market penetration and use of the technology as well as the emergence of gas as an important supplement to oil amongst the conventional sources of energy (ExxonMobil, 2010). See also the discussion on varying estimates and projections for RETs in chapter II of this Report.

15 This estimate includes electricity generated through a variety of renewable resources, including small hydropower (as defined in footnote 6, chapter II), biomass, wind, geothermal and solar (thermal and photovoltaic).

16 For example, for wind and solar energies the thresholds vary for grid parity with conventional sources.

17 In the early 1990s, the German Government began to heavily subsidize the installation of rooftop PV panels as part of its 1,000 Rooftops project. As a result of the initial success, the project was later expanded to a 100,000 Rooftops project (Hader et al., 2011). In Japan, the Ministry of Economy, Trade and Industry initiated a project called the New Sunshine Project in 1993 to develop solar technologies.

18 For example, solar energy is stored in salt and melted salts, but scientists are trying to find ways to use concrete instead. If this technological development materializes, the cost of solar storage would be reduced by half, from €30–40/kWh to below €20/kWh (Hader et al., 2011).

REFERENCES

AGAMA Energy (2003). Employment Potential of Renewable Energy in South Africa. Report prepared for the Sustainable Energy and Climate Change Partnership Johanesburg. Available at http://projects.gibb.co.za/LinkClick.aspx?fileticket=S6HB67wKzQU%3D&tabid=174&mid=797

Amin A (2000). Prototyping structural description using an inductive learningprogram. *IEEE Transactions on Systems, Man, and Cybernetics, Part C: Applications and Reviews*, 30(1): 150-157.

Andersson M and Ejermo O (2004). How does Accessibility to Knowledge Sources affect the Innovativeness of Corporations? - Evidence from Sweden, Royal Institute of Technology, CESIS - Centre of Excellence for Science and Innovation Studies, May.

Archibugi D (2001). Pavitt'S Taxonomy Sixteen Years On: A Review Article. *Economics of Innovation and New Technology*, Economics of Innovation and New Technology, 10(5): 415-425.

Archibugi D and Michie J (1997). Technological globalisation or national systems of innovation? *Futures*, 29(2): 121-137.

Baumol WJ (2003). Welfare Economics and the Theory of the State. In: C K Rowley and F Schneider, eds. *The Encyclopedia of Public Choice*. Boston, MA Springer US.

Bouée C-E, Liu W and Xu A (2011). China - Greed on an unimaginable scale. *Green Growth, Green Profit: How Green Transformation Boosts Business*. Basingstoke Palgrave Macmillan.

Brusoni S and Prencipe A (2001). Managing Knowledge in Loosely Coupled Networks: Exploring the Links between Product and Knowledge Dynamics. *Journal of Management Studies*, 38(7): 1019-1035.

BTM Consult (2009). BTM World Market Update 2009, Copenhagen.

Castelacci F and Archibugi D (2008). The technology clubs: The distribution of knowledge across nations. *Research Policy*, 37(10): 1659–1673

Chandler W and Gwin H (2008). *Financing energy efficiency in China*. Washington, D.C., Carnegie Endowment for International Peace.

Chesbrough HW (2003). Open innovation: the new imperative for creating and profiting from technology. USA, Harvard Business Press.

Ciarli, T., Leoncini, R., Montresor, S. and M. Valente (2008) "Technological Change and the Vertical Organization of Industries", Journal of Evolutionary Economics,18: 367-387.

Chichilnisky G (2009). Financial Innovations & Carbon Markets. *EU Chronicle*, XLVI(3&4).

Cohen WM and Levinthal DA (1990). Absorptive Capacity: A New Perspective on Learning and Innovation. *Administrative Science Quarterly*, 35(1): 128-152.

Cosbey A, Ellis J, Malik M and Mann H (2009). Clean Energy Investment. Project synthesis report, Winnipeg, Manitoba, IISD.

Davis G (2011). The global energy assessment. Presented at Vienna Energy Forum 2011. Vienna. Available at: www.unido.org/fileadmin/user_media/Services/Energy_and_Climate_Change/Renewable_Energy/VEF_2011/Presentations/The%20Global%20Energy%20Assessment%20Presentation.pdf.

Dayo FB (2008). Clean Energy Investment in Nigeria: The domestic context, IISD.

Delucchi MA and Jacobson MZ (2011). Providing all global energy with wind, water, and solar power, Part II: Reliability, system and transmission costs, and policies. *Energy Policy*, 39(3): 1170-1190.

Dosi G, Malerba F, Ramello GB and Silva F (2006). Information, appropriability, and the generation of innovative knowledge four decades after Arrow and Nelson: an introduction. *Industrial and Corporate Change*, Industrial and Corporate Change, 15(6): 891-901.

Economist Intelligence Unit 2007: Industry Forecast Healthcare and Pharmaceuticals India. London. November.

ExxonMobil (2010) The Outlook for Energy: A View to 2030 available at www.exxonmobil.com/energyoutlook

Freeman C (1987). *Technology, policy, and economic performance: lessons from Japan*. London and New York, Pinter Publishers.

Freeman C (1988). Japan: a new national system of innovation? In: G Dosi et al., eds. *Technical Change and Economic Theory*. UK Pinter Pub Ltd.

Gehl Sampath P (2010). *Reconfiguring Global Health Innovation*. Oxon; New York, Routledge.

GTZ (2007). Energy-Policy Market Conditions for Electricity Markets and Renewable Energies: A 23 Country Analyses, Division Environment and Infrastructure TERNA Wind Energy Programme, Eschborn.

Hader M, Hertel G, Körfer-Schün M and Stoppacher J (2011). Renewable energy advancing fast. In: Roland Berger, ed. *Green Growth, Green Profit: How Green Transformation Boosts Business*. Palgrave Macmillan.

Hamilton K (2011). Investing in renewable energy in the MENA region: financier perspectives. EEDP Working Paper, Chatham House, June.

Henzelmann T and Grünenwald S (2011). Green services are the unsung heroes. In: Roland Berger and Von Roland Berger Strategy Consultants GmbH, eds. *Green Growth, Green Profit: How Green Transformation Boosts Business*. Palgrave Macmillan.

Holm D (2005). Renewable Energy Future for the Developing World. ISES White Paper, Friburg, International Solar Energy Society (ISES).

IEA (2009). *World Energy Outlook 2009*. Paris, OECD/IEA.

IEA (2010). *World Energy Outlook 2010*. Paris, OECD/IEA.

IEA (2011). Clean Energy Progress Report, Paris, OECD/IEA.

IPCC (2007). *Climate Change 2007: Mitigation of climate change*. Cambridge, Cambridge University Press.

IPCC (2011). IPCC Special Report on Renewable Energy Sources and Climate Change Mitigation, IPCC.

ISPRE (2009). Research and Development on Renewable Energies: A Global Report on Photovoltaic and Wind Energy, Paris, International Science Panel on Renewabe Energies.

Jaffe A, Lerner J and Stern S, eds. (2005). *Innovation policy and the economy. Volume 5*. Cambridge Massachussets, MIT Press.

Johnson B and Lundvall Bengt-Åke (2003). National Systems of Innovation and Economic Development. In: M Muchie, P Gammeltoft, and Bengt-Ake Lundvall, eds. *Putting Africa First: The Making of African Innovation Systems*. Aalborg Aalborg University Press.

Kalmbach R (2011). India-Cannot afford not to go green. *Green Growth, Green Profit: How Green Transformation Boosts Business*. Palgrave Macmillan.

Krohn S, Morthorst P-E and Awerbuch S (2009). The Economics of Wind Energy, European Wind Energy Association.

Lundvall Bengt-Åke (1985). *Product innovation and user-producer interaction*. Aalborg, Aalborg University Press.

Lundvall Bengt-Ake, ed. (1992). *National Systems of Innovation: Towards a Theory of Innovation and Interactive Learning*. Pinter Pub Ltd.

Malerba F (2002). Sectoral systems of innovation and production. *Research Policy*, 31(2): 247-264.

Malerba F (2004). *Sectoral systems of innovation: concepts, issues and analyses of six major sectors in Europe*. Cambridge UK, Cambridge University Press.

Marsili O and Verspagen B (2002). Technology and the dynamics of industrial structures: an empirical mapping of Dutch manufacturing. *Industrial and Corporate Change*, 11(4): 791 -815.

Martinot E, Chaurey A, Lew D, Moreira JR, et al. (2002). Renewable Energy Markets in Developing Countries. *Annual Review of Energy and the Environment*, 27(1): 309-348.

Martinot E (2003). Renewable energy market indicators and references. Available at: http://www.martinot.info/markets.htm (accessed: 22 August, 2011).

Nelson RR (1987). *Understanding Technical Change As an Evolutionary Process*. Elsevier Science Ltd.

Nelson RR (1993). *National Innovation Systems: A Comparative Analysis*. Oxford University Press, USA.

Nelson RR and Winter SG (1982). *An evolutionary theory of economic change*. Harvard University Press.

Neuhoff K (2008). Learning by Doing with Constrained Growth Rates:An Application to Energy Technology Policy. *The Energy Journal*, 29(01)

Ocampo JA and Vos R (2008). *Uneven economic development*. United Nations Publications.

OECD (1998). The Emerging Digital Economy Report, Paris, Organization for Economic Cooperation and Development.

OECD (2010). Interim Report of the Green Growth Strategy: Implementing our Commitment for a Sustainable Future, Paris, OECD.

Point Carbon (2007). Carbon 2007. A new climate for carbon trading.

Practical Action (2010). Poor people's energy outlook 2010, Rugby, UK.

Ramachandran V, Gelb A and Shah M (2009). Africa's Private Sector: What's Wrong with the Business Environment and What to Do About It – Brief.

Reers J, Benecchi A and Koper S (2011). The United States - A chance to reinvent itself. *Green Growth, Green Profit: How Green Transformation Boosts Business*. Palgrave Macmillan.

Reinert ES (2007). *How Rich Countries Got Rich and Why Poor Countries Stay Poor*. Constable.

REN21 (2011). Renewables 2011 Global Status Report, Paris, REN21 Secretariat.

Robertson PL and Langlois RN (1995). Innovation, networks, and vertical integration. *Research Policy*, Research Policy, 24(4): 543-562.

Rosenberg N (1982). *Inside the Black Box: Technology and Economics*. Cambridge UK; New York, Cambridge University Press.

Sarker AR and Alam K (2010). Nexus between Electricity Generation and Economic Growth in Bangladesh. *Asian Social Science*, 6(12): 16-22.

Siegel J (2006). Financing Renewable Energy in Developing Countries2006. Available at: http://apps.americanbar.org/environ/committees/renewableenergy/teleconarchives/041906/Siegel_Presentation.pdf.

Skoglund A, Leijon M, Rehn A, Lindahl M, et al. (2010). On the physics of power, energy and economics of renewable electric energy sources - Part II. *Renewable Energy*, 35(8): 1735-1740.

Stäglich J, Lorkowski J and Thewissen C (2011). Electric mobility comes of age. In: Von Roland Berger Strategy Consultants GmbH, ed. *Green Growth, Green Profit: How Green Transformation Boosts Business*. Palgrave Macmillan.

UNCTAD (2005). *World Investment Report 2005: Transnational corporations and the internationalization of R&D*. United Nations publication, sales No. E.05.II.D.10. New York and Geneva, United Nations.

UNCTAD (2007). *Regional cooperation for development*. New York and Geneva, United Nations.

UNCTAD (2010). *World Investment Report: Investing in a Low-carbon Economy*. United Nations publication, sales no. E.10.II.D.2. New York and Geneva, United Nations.

UN/DESA (2009). A Global Green New Deal for Climate, Energy, and Development. New York, United Nations.

UN/DESA (2011). *World economic and social survey 2011: the great green technological transformation.* New York, United Nations Department of Economic and Social Affairs.

UNEP (2011). *Towards a Green Economy: Pathways to Sustainable Development and Poverty Eradication - A Synthesis for Policy Makers*, www.unep.org/greeneconomy.

UNFCCC (2010). Outcome of the work of the Ad Hoc Working Group on long-term Cooperative Action under the Convention.

4

INTERNATIONAL POLICY CHALLENGES FOR ACQUISITION, USE AND DEVELOPMENT OF RENEWABLE ENERGY TECHNOLOGIES

CHAPTER IV

INTERNATIONAL POLICY CHALLENGES FOR ACQUISITION, USE AND DEVELOPMENT OF RENEWABLE ENERGY TECHNOLOGIES

A. INTRODUCTION

International discussions and negotiations on climate change and the green economy have gained momentum in recent years. A major area under consideration relates to environmentally sustainable technologies, or low-carbon, "clean" technologies, as a means of contributing to climate change mitigation and adaptation globally.[1] This is a very important global goal, which will serve the needs of developing countries, in particular, given the evidence that climate change will have disproportionately damaging impacts on those countries. However, even as efforts are made to mitigate climate change, there needs to be an equally important focus on eliminating energy poverty in developing countries, not only to improve people's living conditions, but also to boost economic development, as earlier chapters of this *TIR* have stressed.

This chapter calls for a repositioning of issues within the international agenda, whereby the obligations of countries to mitigate climate change are framed in terms of creating development opportunities for all in an environmentally sustainable manner. Central to this repositioning is the triangular relationship between equity, development and environment. From this perspective, recognition of the right of all people worldwide to access energy services (as discussed in chapter I) is long overdue and needs to be addressed. Developing countries, especially the least developed, have experienced a particularly large share of natural disasters, such as hurricanes, tornados, droughts and flooding, as a result of changing climatic conditions. According to recent estimates, 98 per cent of those seriously affected by natural disasters between 2000 and 2004 and 99 per cent of all casualties of natural disasters in 2008 lived in developing countries (Tan, 2010; Global Humanitarian Forum, 2009; UNDP, 2007; and UN/DESA, 2009a), particularly in Africa and South Asia where the world's poorest people live. These disasters have not only caused food shortages; in many instances, they have also ruined the livelihoods of large numbers of people already living in extreme poverty. Consequently, the heightened economic insecurity caused by climatic events has been borne disproportionately by developing countries.

As efforts are made to mitigate climate change, there needs to be an equally important focus on eliminating energy poverty in developing countries.

A repositioning also implies focusing on issues of finance and technology transfer and acquisition for developing countries, especially in the context of RETs. These issues may be considered within ongoing discussions on financing for climate change adaptation, but they may also require separate and newer initiatives that focus particular attention on enabling the economic development of countries and people. Specifically, how can adequate resources be mobilized to ensure that people living in developing countries secure access to energy and employment opportunities? Efficiency in meeting the energy needs of developing countries requires use of the most efficient technologies available worldwide (Birdsall and Subramanian, 2009).

Such a shift in positioning implies focusing on issues of finance and technology transfer and acquisition for developing countries, especially in the context of RETs.

This chapter argues for the need for international support to complement national frameworks that seek to promote technology and innovation capacity in RETs. It analyses three important policy challenges related to climate change and RETs: (i) international resource mobilization for RETs financing, (ii) greater access to technology through technology transfer and the greater use of flexibilities in the intellectual property (IP) regime, and (iii) promoting technological learning and wider use of RETs through the green economy and the Rio+20 framework. These issues have been and remain central to all debates and decisions of the UNFCCC and the Kyoto Protocol. Much of these discussions refer to environmentally sustainable, or clean, technologies,[2] of which RETs form a subset. The chapter examines financing, technology transfer and IP issues that are being discussed in international negotiations and in debates to the extent that they apply to RETs. By highlighting the key international developments, often conflicting policy developments[3] and the main hurdles that remain to be overcome, the chapter calls for the international discourse to consider the needs of developing countries for science, technology and innovation of RETs. In this context, it makes proposals for greater international support to developing countries, including an international innovation network of RETS for LDCs, global and regional research funds for RETs deployment and demonstration, an international RETs technology transfer fund and an international RETs training platform.

The role of RETs in complementing and eventually even replacing existing energy sources worldwide will remain just rhetoric...

... if they are prohibitively expensive.

B. INTERNATIONAL RESOURCE MOBILIZATION AND PUBLIC FINANCING OF RETs

The role of RETs in complementing and eventually even replacing existing energy sources worldwide will remain just rhetoric if they are prohibitively expensive. Finance has been at the forefront of all issues in international discussions on climate change mitigation. This is largely because of the dauntingly large amounts of investments in RETs that are needed if the world is to avoid dangerous anthropogenic climate change. An important forum where this issue of financing is being discussed is the UNFCCC. Although these discussions refer to environmentally sustainable, or clean, technologies,[4] of which RETs form a subset, these discussions offer an important basis to analyse the key issues relating to international resource mobilization for RETs.

1. Financing within the climate change framework

Several proposals have been made concerning the sharing of the burden of climate change mitigation (box 4.1). The UNFCCC[5] stipulates that developed countries should ensure the availability of "new and additional financial resources" to meet the "agreed full costs" involved in enabling developing countries to meet their national commitment requirements under Article 12 of the Convention. Article 4(3) calls on developed countries to provide "such financial resources, including the transfer of technology" to all developing countries to meet "the agreed full incremental costs" of implementing mitigation and adaptation actions and other commitments identified in Article 4(1), including reporting of emissions and carbon sink removals, integration ²čof climate change considerations into national policies, education, training and public awareness, and research on climate change. Additionally, the UNFCCC requires commitments from developed countries to finance adaptation costs in developing countries.[6]

A number of estimates have been produced that try to quantify the challenge of adaptation and climate change mitigation (see table 4.1 for summaries of the major estimates). All of them consider slightly different categories of investments that will be needed in the immediate or medium term. The International Energy Agency (IEA, 2010a) estimate covers only electricity generation technologies, and therefore excludes investment in transport fuels and heating technologies. The UNFCCC (2008) estimates cover only power generation, which includes carbon capture and storage (CCS), nuclear and large-scale hydro. While all the estimates are indicative, the definitions of technology and the broad goals

Box 4.1: Kyoto Protocol, emissions control and burden sharing

Climate change has been touted as the greatest market failure in the world. The Kyoto Protocol and important reports on the topic, such as the *Stern Report on the Economics of Climate Change of 2007* and UNDP's *Human Development Report 2008*, have advocated allocating the future burden of emission reductions according to an 80-20 formula whereby the rich countries agree to cut emissions by 80 per cent by 2050 from their 1990 levels and poor countries by 20 per cent. From a historical perspective, it has been suggested that it is mainly emissions from today's developed countries during their process of industrialization that have contributed to the level of greenhouse gases (GHGs) in the world today. However, China has now replaced the United States as the world's largest emitter of GHGs, and India and China together host over a third of the world's population. If these and other countries were to pursue industrialization using the same conventional technologies and energy sources that developed countries used during their process of industrialization, the effects on climate change, with all their ramifications, would be unthinkable, as chapter I has pointed out. Yet, limiting developing countries' choices of technologies and energies to more climate-friendly ones, which require greater investments and costs, could imply their having to forsake development opportunities. The debates on burden sharing by different countries therefore tend to focus on their varying capabilities and contributions to the past and current levels of GHG emissions as well as the opportunity costs of shifting to a low-carbon, high-growth mode of development.

Five main proposals have emerged over time (Mattoo and Subramanian, 2010). The equal per capita emissions proposal is based on the premise that, regardless of all past and future responsibilities, every country should be treated alike in assessing its right to emit GHGs. A second proposal, based on historical responsibility, suggests that the allocation of rights to all future emissions should be inversely linked to the past emission records of countries. A third proposal is based on ability to pay, and links payments for climate change mitigation and adaptation to poverty criteria. There are several versions of this proposal, the most extreme version proposing that below a particular level of income, individuals or countries themselves will have no obligation to pay. A fourth proposal seeks to preserve future development activities by allocating sufficient carbon allowances to countries that are currently poor and have not made enough use of their carbon allowances for development purposes. A fifth proposal relates to the distribution of adjustment costs. The 80-20 formula by Stern (2007) and UNDP (2008), also fall in this category.

It is not clear how these proposals fit into the broadly accepted and negotiated "common-but-differentiated" obligations of the UNFCCC. Indeed, the issue of burden sharing has proved to be problematic.

Source: UNCTAD.

Table 4.1: Estimates of RETs investments needed for climate change adaptation and mitigation

Source	Publication	Annual investment needed	Purpose of investments
IEA	*Energy Technology Perspectives 2000*	\$300–\$400 billion between 2010 and 2020, and up to \$750 billion by 2030	Low-carbon energy technologies needed to achieve IEA's BLUE Map scenario (related CO_2 emissions fall by half between 2005 and 2050)
IEA	*World Energy Outlook 2010*	Average of \$316 billion between 2010 and 2035	Electricity generation only; needed to reach IEA's 450 scenario (i.e. atmospheric GHGs at 450 particles per million (ppm) CO_2 equivalent, or an average temperature rise of 2°C)
UNFCCC	*Investment and Financial Flows to Address Climate Change: An Update (2008)*	\$148.5 billion by 2030	Needed to reduce GHG emissions by 25 per cent below 2000 levels. Includes CCS, nuclear and large hydro
UNEP SEFI/ Bloomberg New Energy Finance	*Global Trends in Sustainable Energy Investment 2010*	\$500 billion by 2030	To reduce GHG emissions from 42 Gt to 39 Gt by 2030. Excludes large (>50 MW) hydro

Source: UNCTAD, based on Cosbey and Savage (2011).

assumed in the IEA (2000) are probably the most relevant. The proposal to halve energy-related emissions by 2050 corresponds roughly to the minimum mitigation levels deemed necessary by the IPCC, and the definition of low-carbon energy technologies corresponds well to the scope of this *TIR*, which covers RETs. The IEA's estimates for the level of investments needed are lower than the other estimates in the medium term, at \$300–\$400 billion per annum up to 2020, but rise thereafter to reach \$750 billion by 2030.

All these analyses examine the conditions broadly necessary to bring about technological transformation. As such, they

Table 4.2: Multilateral and bilateral funding for low-carbon technologies

	Fund	Total amount ($ million)
Major multilateral initiatives		
World Bank	Climate Investment Funds	6 100
	Clean Technology Fund	4 700
	Strategic Climate Fund	1 400
International Finance Corp.	Sustainable Energy and Water	2 000
GEF-4 (various, incl. land-use change and forestry)		1 400
Asian Development Bank	Climate Change Fund	40
	Clean Energy Financing Partnership Facility	90
	Poverty and Environment Facility	4
European Development Bank	Multilateral Carbon Credit Fund	276
Subtotal		**16 009**
Major bilateral initiatives		
Japan	Hatoyama Initiative[a]	15 000
Netherlands	Development Cooperation	725
Australia	International Forest Carbon Initiative	132
United Kingdom	Environmental Transformation Fund[a]	1 182
Norway	Climate Forest Initiative	2 250
Germany	International Climate Initiative	764
European Commission	Global Climate Change Alliance	76
Spain	MDG Achievement Fund	92
Subtotal		**20 221**
Total		**36 230**

Source: IEA (2010b).

[a] Also includes funding for adaptation, which has little if any relevance for RETs.

include investments throughout the innovation chain, from R&D through demonstration and commercialization to dissemination and deployment of RETs (figure 4.1).

This raises questions about the capacity of public finance to support the rapid and widespread deployment of RETs as part of adaptation efforts and the role of international support.

This raises questions about the capacity of public finance to support the rapid and widespread deployment of RETs as part of adaptation efforts and the role of international support. There are a number of known sources of finance at the multilateral and regional levels (table 4.2). Notable examples include the World Bank's Climate Investment Funds and, specifically, the Clean Technology Fund, with commitments of $4.5 billion until 2010. As of November 2010 the Clean Technology Fund had approved $2.4 billion to support large-scale renewable deployment in 14 middle-income developing countries (Algeria, Egypt, Indonesia, Jordan, Kazakhstan, Mexico, Morocco, the Philippines, South Africa, Thailand, Tunisia, Turkey, Ukraine and Viet Nam), and through its Scaling Up Renewable Energy in Low Income Countries Program, it had planned to provide support for renewables in an additional six pilot low-income countries (World Bank, 2010).

Table 4.2 provides a list of multilateral and bilateral funding programmes, but it is by no means exhaustive. For instance, it does not include the newly announced UNFCCC Green Climate Fund, since the details of this Fund are not yet known. Neither does it include the $30 billion fast-start financing between 2010 and 2012, and the target to mobilize $100 billion per year by 2020 (UNFCCC, 2010), which was announced as part of the Copenhagen Accord (at least some of which will certainly be administered by the Green Climate Fund). This funding was signed into UNFCCC commitments as part of the Cancun Agreements in December 2010.

Figure 4.1: Funding arrangements of the UNFCCC

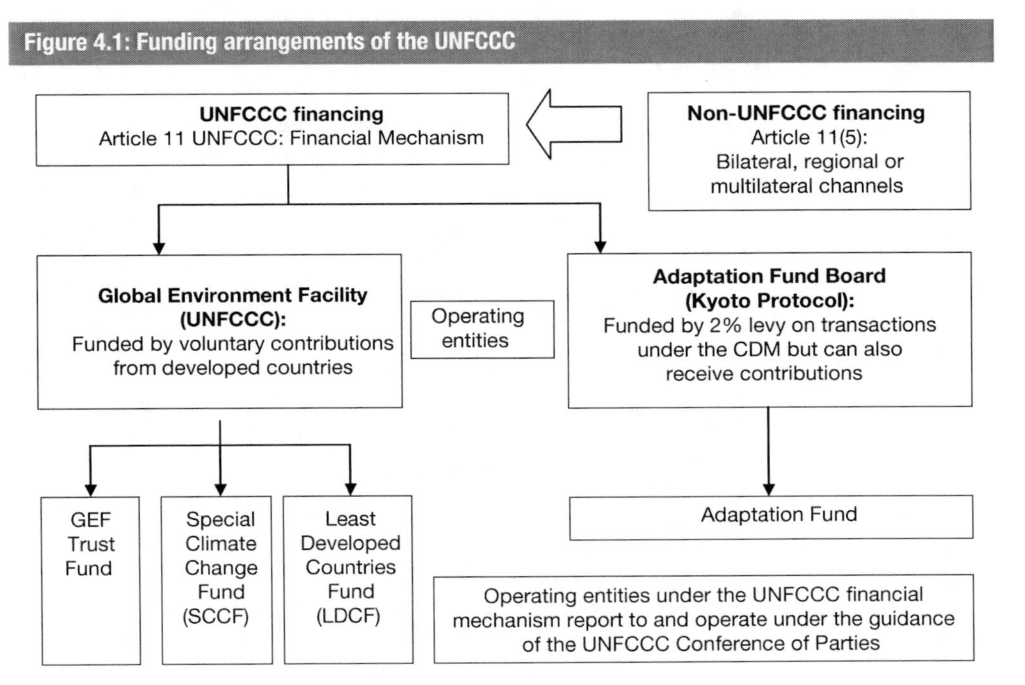

Source: Reproduced from Tan (2010).

A number of caveats with respect to these funding figures need to be considered before they can be accurately compared with the estimates of need shown in table 4.1. First, all are multi-year commitments, whereas the figures in table 4.1 are annual requirements. Several of the largest sources of funds in table 4.2 are slated to cover both mitigation *and* adaptation. Finally, while the commitments made under the Cancun Agreements are valuable, the long-term $100 billion dollar benchmark is a commitment to *mobilize* that amount of money from both the private and public sector, and there is no clear idea so far as to the respective shares of the two sectors. Also, some of these funds are not yet available. Taking all these caveats together, the total amount of annual funding for RETs from public sources is likely to be about $5 billion from the known sources in table 4.2. If the full funding levels under the Cancun Agreement are reached, they will contribute some percentage of $100 billion of government support for RETs, which is only a part of the $100 billion dollar commitment to annual funding targeted to be achieved by the year 2020. This falls far short of the estimates of the needed investment, which range from a low of about $150 billion to a high of $750 billion by 2030.

If the full funding levels under the Cancun Agreement are reached...

The Conference of the Parties (COP) to the UNFCCC initiated a funding mechanism through the Global Environment Facility (GEF) in 1998 and also created the Adaptation Fund Board in 2008, which deals with climate change mitigation and adaptation, including, *inter alia,* RETs. The GEF operates three trust funds established under the Convention – the GEF Trust Fund, the Strategic Climate Change Fund (SCCF) and the Least Developed Countries Fund (LDCF) – while the Adaptation Board operates the Adaptation Fund (figure 4.1). The first three funds rely on voluntary contributions from all UNFCCC parties, both developed and developing countries, while the Adaptation Fund is funded by a 2 per cent levy on transactions under the Kyoto Protocol's Clean Development Mechanism (CDM) (box 4.3).

...this falls far short of the estimates of the needed investment, which range from a low of about $150 billion to a high of $750 billion by 2030.

2. Other sources of finance

Figure 4.2 shows how the various sources of finance typically contribute at different

At the early stages of research, funding is almost entirely public.

As the innovation progresses and becomes more demonstrably sound, it can attract funding from public equity markets.

stages of the innovation chain (i.e. the sequence of stages through which any innovation must pass before it becomes widely disseminated). That chain starts with R&D, then moves to manufacturing scale-up (i.e. developing scaled up commercial processes and building demonstration projects) and finally ends with roll-out, which involves investment in large-scale deployment and dissemination of the technology. At the early stages of research, funding is almost entirely public, though some may also be undertaken in private sector research divisions. When a concept reaches the development stage, while it may still benefit from government support, it becomes interesting enough to attract venture capital and private equity – sources of finance that tolerate higher risks (and demand commensurately higher returns). As the innovation progresses and becomes more demonstrably sound, it can attract funding from public equity markets, or it may become the object of a merger or acquisition. When the concept appears viable on a commercial scale, debt-financing becomes an option. It should be noted that figure 4.2 does not show the relative scale of funding needs, which become progressively greater along the innovation chain to the extent that it would be highly unlikely that a government alone would engage in asset finance for commercial roll-out (funding the development of a wind farm, for example).

Currently, there are a number of multilateral programmes and funds – typically aiming to help achieve the goals of climate change mitigation – that might support the difference in costs between conventional and renewable new capacity in large-scale generation scenarios where RETs are more costly than conventional alternatives. At the same time, a number of national and international grassroots agencies are working to promote RETs as solutions to energy poverty, and many of them are supported by multilateral funding agencies.

Medium-scale projects that represent technological improvements on existing RETs, or mini-grid-based applications

Figure 4.2: Sustainable energy financing along the innovation chain

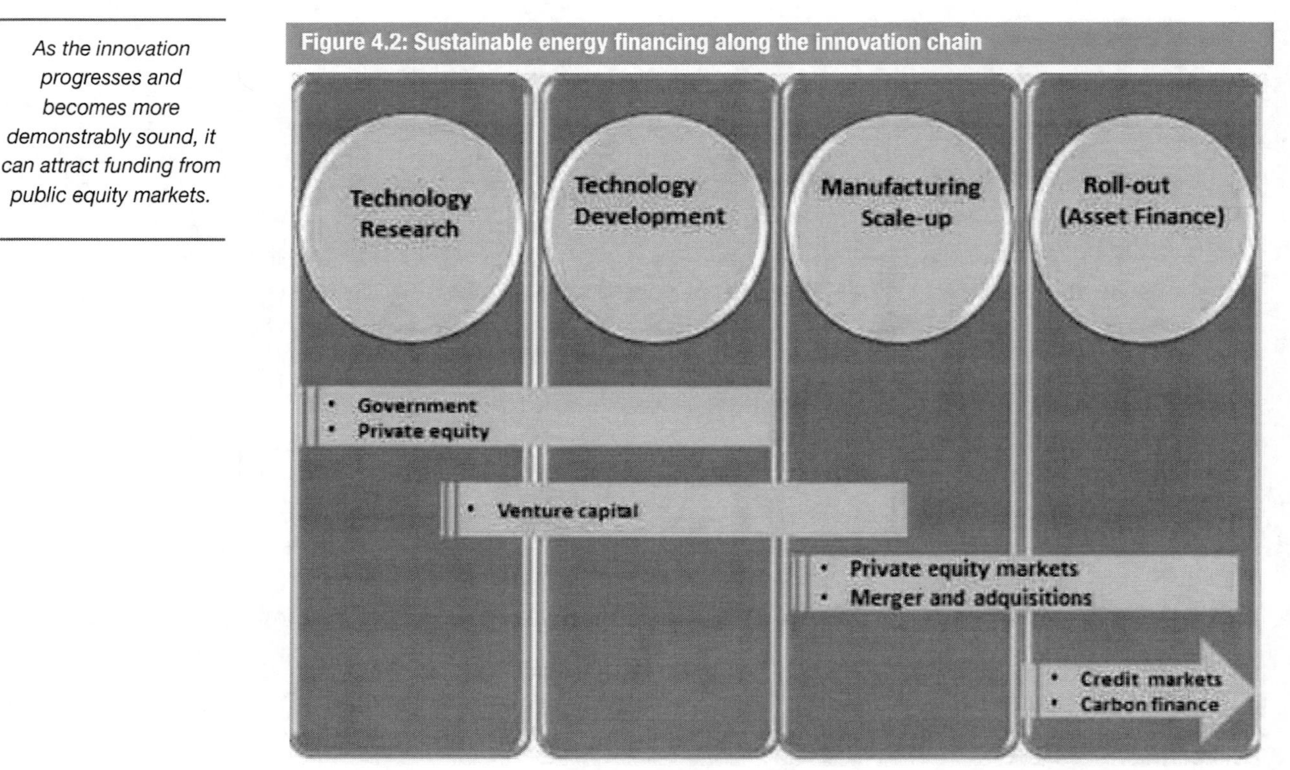

Source: UNCTAD, adapted from UNEP and Bloomberg (2010).

using RETs which can be combined with conventional energy sources, offer innovative niches for firms in developing countries to learn through adaptation of existing technologies and innovations. However, in general, developing country firms initiatives lack access to the extensive financing and venture capital infrastructure needed for such entrepreneurial initiatives. It also seems evident that while in the near future governmental and international support will play a critical role in expanding the use, adaptation and innovation in RETs across countries, government support can only constitute a small proportion of the total amount of investments needed for this purpose. In the long term, the major source of investment in RETs needs to be non-governmental, mostly in the form of asset finance (from utilities and energy groups from their own balance sheets, or from debt or equity finance) and private equity and venture capital finance, or contributions from homeowners and small installations for small distributed capacity. Therefore, the best way to position such international support is to ensure that it plays a catalytic role to help facilitate the much larger flows of private investment needed for widespread dissemination and building of technological capabilities in RETs in developing countries.

3. International support for financing of RETs: Outstanding issues

A macroeconomic climate that supports greater investment in RETs and their use in developing countries is critical today. In response to the global financial and economic crisis, many countries initiated stimulus packages which included efforts to build capacity in areas of the green economy that display the greatest growth potential. Approximately $188 billion of "green" stimulus spending began to be disbursed globally in 2009, much of it focusing on renewable energy technologies (UNEP and Bloomberg, 2010). The United States made the largest commitment, of $66.6 billion, followed by China and the Republic of Korea with $46.9 billion and $24.7 billion respectively. The only other developing country in the top 10 pledges was Brazil with $2.3 billion (Ibid.). China's "green push" amounted to over 34 per cent of its stimulus spending, and it targeted areas such as green transportation, smart grids, low-carbon vehicles, and advanced waste and water infrastructure (Robins, Clover and Singh, 2009). While not all of the green spending was targeted at REs, it is estimated that RETs received support amounting to about $57 billion that year (IEA, 2010a). As chapter III has pointed out, by early 2010, over 100 countries – developed and developing – were providing policy support to promote the use and dissemination of RETs. No doubt, the general trend is towards policies that simultaneously aim at securing environmental benefits through increased use of RETs, development benefits through increased energy provision, and economic benefits by increasing domestic capacity in areas that show growth potential.

Firms in developing countries lack access to the extensive financing and venture capital infrastructure needed for such entrepreneurial initiatives.

However, such ongoing efforts in developing countries would be better served if outstanding issues relating to international financial support for RETs could be urgently resolved with the aim of promoting greater innovation, production and use of such technologies. The major issues are summarized below.

(i) At present, international financing of clean technologies, which is largely multilateral, is highly fragmented, uncoordinated and lacks transparency. Figuring largely within the climate change framework, its high degree of stratification, with multiple funds and complex financing arrangements, results in considerable overlap. It is also highly inadequate to meet total funding requirements for climate change mitigation and adaptation (Tan, 2010; UN/DESA, 2009b). While such financing may partly cover RETs, additional international funding for RETs is required as a priority. IRENA estimates that Africa alone needs

International financing of clean technologies, which is largely multilateral, is highly fragmented, uncoordinated and lacks transparency.

$40 billion per annum for its power sector. The requirements of other developing regions are likely to be similar, if not higher.

(ii) There is a large gap between the financial needs of countries and what is provided for RETs. It is therefore necessary to explore what additional measures are needed to address these shortcomings.

There is a large gap between the financial needs of countries and what is provided for RETs.

(iii) Coordination of funding sources with the aim of mainstreaming RETs into national energy systems globally should be an important aspect of climate change mitigation efforts. This would not only lead to the development of more efficient energy systems globally; it would also ensure that the financing results in more technological progress towards newer and/or more cost-effective RETs.

(iv) International policy support for RETs by setting an international target (specifying the extent to which RETs would form part of the global energy system by a particular deadline) would be an important policy signal to national governments, as well as private investors. Given the uncertainties surrounding the development of these technologies, a target-based policy signal (as opposed to stop-and-go policy signals) would be important for stimulating investments.

International financing for RETs should be coordinated with national aspirations for RETs development and expansion.

(v) International financing for RETs should be coordinated with national aspirations for RETs development and expansion. As chapter III points out, a number of developing countries have enacted policy targets to provide greater access to energy through RETs expansion. International financing of RETs should be coordinated with such policy goals of countries and mainstreamed into development cooperation programmes.

C. TECHNOLOGY TRANSFER, INTELLECTUAL PROPERTY AND ACCESS TO TECHNOLOGIES

Although technology transfer has been a key issue in international forums and an intrinsic part of treaty texts since the 1970s, there is no single, widely accepted definition of this term (see, for example, Patel, Roffe and Yusuf, 2000; and Maskus and Reichman, 2005). At a general level, the growing technological divergence between developing countries since the 1970s[7] and the gradual technological downgrading witnessed among the LDCs (see UNCTAD, 2010) have prompted discussions on how technology transfer to developing countries and LDCs could be promoted. Article 66.2 of the Agreement on Trade-Related Aspects of Intellectual Property Rights (TRIPS) of the World Trade Organization (WTO) embodies a legal obligation on the part of developed-country members to provide specific incentives for promoting technology transfer to LDCs (Correa, 2005; and box 4.2).

Generally, the transfer of technology can occur on a day-to-day basis, including through informal channels such as the circulation and transfer of educational materials, skills accumulation through employment of local people in international firms, trade fairs, and general education and training. Some of the more formal means of technology transfer include joint ventures that promote the sharing of know-how, training of local personnel, or simply employment of local staff; technical assistance programmes and other forms of aid; research collaborations in the public and private sectors; subcontracting agreements and technology licensing contracts (see, for example, UNCTAD, 2007 and 2010). In all its forms, technology transfer is central to accessing relevant knowledge through the transfer of not only codified information (in the form of blueprints, technology and equipment), but also intangible

Box 4.2: Technology transfer and Article 66.2 of the WTO Agreement on Trade-Related Aspects of Intellectual Property Rights

The legal obligation for technology transfer contained in Article 66.2 of the TRIPS Agreement is reflected in the following provision: "Developed country members shall provide incentives to enterprises and institutions in their territories for the purpose of promoting and encouraging technology transfer to least developed country members in order to enable them to create a sound and viable technological base." Clearly, the intent of this provision is to encourage the transfer of technology to LDCs members of the WTO. To what extent this has materialized in practice is a matter of intense dispute. One of the only reviews on the topic examines whether Art. 66.2 has resulted in an increase in business between developed countries and LDCs (Moon, 2008). Based on country self-reports to the TRIPS Council between 1999 and 2007, and focusing mainly on the public policies and programmes that developed countries undertake to encourage their organizations/enterprises to engage in such technology transfer, the study concludes that the absence of a clear definition of key terms such as "technology transfer" and "developed country" renders it difficult to determine which members are obligated to provide incentives, in what form and towards what ends. Since many countries did not submit reports regularly to the Council, and those that submitted did so irregularly, the review concludes that of 292 programmes and policies reported, only 31 per cent specifically targeted LDC members of the WTO. Of these, approximately a third of the programmes that did target LDCs did not specifically promote technology transfer. Thus, of the 292 programmes, only 22 per cent involved technology transfer specifically targeted at LDC members (Moon, 2008). At the April 2010 session of the Committee on Development and Intellectual Property (CDIP) in World Intellectual Property Organization (WIPO), the group of like-minded developing countries[a] called for a study on the extent to which TRIPS Article 66.2 had been fulfilled. However, developed countries contend that the article reaches beyond the mandate of the WIPO into business, trade, financial and other areas.

Source: UNCTAD.

[a] The like-minded countries are mainly the African Group, the Arab Group, Brazil and India.

know-how, which is an essential component to enable developing-country recipients to absorb, apply and use the technology for various industrial purposes (Arora and Gambardella, 2004).

In the specific context of technology transfer, whatever the channel, the acquisition of information concerning the technology is only one part of the whole process of transfer. The process of learning how to use and maintain the technology is just as important, as is the capacity to adapt the acquired technology to local conditions. Such adaptation may ultimately lead to the development of new applications of the transferred technology. Thus, successful technology transfer goes through the phases of acquiring information, assimilation and absorption of technological knowledge, adaptation to local conditions, absorption of subsequent improvements and the dissemination of the transferred knowledge. These phases jointly account for the complex process of technology transfer.

Mounting evidence on the role of tacit knowledge and the presence of human skills to absorb technologies (learning by doing, and incremental innovation in developing countries) shows that the eventual success of technology transfer depends on the ability of local actors to absorb and effectively use and innovate imported technologies. In sum, successful transfer of technologies depends more on local capabilities to absorb than on the technologies themselves. Narratives about industrial policy are replete with examples of countries that managed to build sectors primarily on the basis of consistent investments in technological capabilities without large-scale transfers of technology. Local capacity to develop locally appropriate productive technologies (and to adapt existing technologies to local conditions) is an essential adjunct to effective policies relating to technology transfer and adaptation. For RETs as for any other sector, local capacity in a variety of areas is important for adaptation, dissemination and use as much as for innovation, as discussed in chapter III.

International support to promote access to existing technologies and related know-how should complement national efforts to boost absorptive capacity by supplying much-needed technological know-how.

The process of learning how to use and maintain the technology is just as important, as is the capacity to adapt the acquired technology to local conditions.

1. Technology transfer issues within the climate change framework

A key issue that has emerged in the climate change negotiations relates to greater access to clean technologies. Indeed there was a similar emphasis when the issue of transfer of environmentally sound technologies was first raised in chapter 34 of Agenda 21 of the 1992 United Nations Conference on Environment and Development.[8]

Article 4.5 of the draft UNFCCC has emerged as the lynchpin of the debate.

Of the various provisions, Article 4.5 of the UNFCCC has emerged as the lynchpin of the debate. Article 4.5 of the draft UNFCCC calls on developed countries to take adequate steps to promote the transfer of technology to developing countries. It states: "The developed countries and other developed countries in Annex II shall take all practicable steps to promote, facilitate and finance, as appropriate, the transfer of or access to environmentally sound technologies and know-how to other Parties, particularly developing country Parties, to enable them to implement the provisions of the Convention." Furthermore, Article 4.7 recognizes that implementation of commitments by developing-country parties to the Convention will depend upon their receiving appropriate transfers of technology. It states: "The extent to which developing country Parties will effectively implement their commitments under the Convention will depend on the effective implementation by developed country Parties of their commitments related to financial resources and transfer of technology."

The following five themes were identified for further discussion by the COP at the seventh session of the UNFCCC in 2001:

(i) *Technology needs and needs assessment*, comprising a set of country-level activities that identify mitigation and adaptation technology as priorities;

(ii) *Technology information*, comprising all hardware and software components that could facilitate the flow of information aimed at enhancing the transfer of environmentally sound technologies;

(iii) *Enabling environments*, consisting of government actions, including policies on fair trade, removal of technical, legal and administrative barriers to the transfer of technology, economic policies, regulatory frameworks, and greater accountability and transparency;

(iv) *Capacity-building*, which refers to processes to build, strengthen and promote existing scientific and technical capabilities and institutions in developing countries with a view to enhancing their capacity to absorb environmentally sound technologies; and

(v) *Technology transfer mechanisms*, to support financial, institutional and methodological activities, with a view to enhancing coordination of stakeholders in different regions and countries, engaging in technological cooperation and partnerships at all levels, and developing new projects to support these goals.

These five core themes cover key aspects relating to technology transfer, including financing, supportive regulatory frameworks, institution building and developing greater absorptive capacity for environmentally sound technologies in developing countries. A closer look at the proceedings of the thirteenth session of the COP to the UNFCCC in 2007 shows a clear consensus that technology transfer is central to the implementation of the Convention beyond 2012. In order to follow through on the five issues identified in the seventh session of the COP in 2001, an expert group on technology transfer was established, and the Bali Action Plan of 2007 called for: "enhanced action on technology development and transfer to support action on mitigation and adaptation, including, *inter alia*, consideration of effective mechanisms and enhanced means for the removal of obstacles to, and provision of financial and other incentives for, scaling

Box 4.3: The Clean Development Mechanism and technology transfer

Article 12 of the Kyoto Protocol defines the CDM as a market-based mechanism set up as an incentive for the financing and diffusion to developing countries of emission-reducing technologies by the private sector. Several studies have sought to evaluate CDM's role in promoting implementation of the Protocol (see, for example, Dechezleprêtre, Glachant and Ménière, 2008; and Seres, Haites and Murphy, 2009). The CDM was a means by which Annex I countries could initiate projects that would result in "certified emission reductions" (CERs) in non-Annex I countries, thereby helping to mitigate climate change. According to the conceptual logic of this mechanism, Annex I countries would be able to count the certificates from their projects as part of their quantified emission targets under Article 3 of the Protocol. In essence, it was a way for developed countries to offset their carbon emissions by supporting emission reduction projects in developing countries that would help them to purchase carbon credits. There is considerable disagreement on how and to what extent mechanisms such as CDM can help raise finances for climate change mitigation (and technology transfer) due to the high volatility of carbon markets. A second area of discord relating to the CDM is the financial relationship between the CDM and the Adaptation Fund which is meant to support climatically vulnerable countries under Article 12.8 of the Kyoto Protocol.

Source: UNCTAD.

up of the development and technology to developing country Parties in order to promote access to affordable environmentally sound technologies (ESTs)".

Developing countries (G-77+China) proposed the establishment of an institutional mechanism on technology transfer and the creation of a new Multilateral Climate Technology Fund in 2008 at the UNFCCC. The dominant view was that IPRs had to be addressed in a systematic and cross-cutting manner to enhance access to environmentally sustainable technologies.

To this end, the Cancun Climate Change Conference in 2010 proposed a new Technology Mechanism which could help enhance the technological capacity of countries to absorb and utilize RETs. The Mechanism is intended to bolster international support for technology development and transfer, particularly to developing countries, in support of climate change mitigation and adaptation. However, the financial and institutional aspects of the Mechanism still need to be worked out, especially its two components: the Technology Executive Committee and the Climate Technology Centre and Network. Some earlier efforts, such as the CDM (box 4.3), may also need to be reconsidered and enhanced along with private sector initiatives to facilitate the greater diffusion of RETs.

2. Intellectual property rights and RETs

Over the past decade, development scholars have tried to reconcile the universal standards of IPR protection that the TRIPS Agreement calls for, with the technological realities of developing countries and LDCs. As noted in the relevant economic literature, the extension of IPRs entails costs of various kinds to developing countries, especially to those countries that are not particularly technologically advanced (Maskus, 2000).[10] At the same time, three broad indirect benefits from extending IPRs have been highlighted for a large number of developing and LDCs where R&D capabilities are low: (i) higher foreign direct investment (FDI), technology transfer, licensing and technology sourcing of value-added goods through foreign subsidiaries with potential positive impacts on domestic learning; (ii) greater innovative activities from access to patent disclosures and technologies; and (iii) competitive returns to innovative firms in developing countries from stronger IPRs and less legal uncertainty (Lai and Qiu, 2003). It has been argued that all of these (especially increased technology transfer and FDI) would help developing countries catch up, and even leapfrog.

Ways and means to address these issues and to finance innovation of relevance to the poorer countries remain controversial. The policy debates on IPR issues

There is increasing patent activity in many clean energy technologies by the OECD countries.

tend to emphasize the safeguards (generally referred to as "flexibilities" in the TRIPS Agreement) contained in the global IPR regime – notably parallel imports and compulsory licensing – which are limited in scope.[11] Moreover, many countries have, to varying degrees, forgone these flexibilities through "TRIPs-plus" regimes adopted by major technology exporters. The debates have focused on certain key areas of public interest such as health,[12] agriculture,[13] and, more recently, climate change.[14] The key issues and way ahead for developing countries with regard to the use, adaptation and innovation of climate-friendly RETs are discussed at length here.

a. The barrier versus incentive arguments

Many of the RETs needed to alleviate energy poverty in developing countries are off-patent (IPCC, 2007). On the issue of new inventions, a recent joint study by UNEP, EPO and ICTSD (2010) points out that there is increasing patent activity in many clean energy technologies by the OECD countries, and that most of the applications for these patents are being filed in developed countries and China. Noting the strategic international trade motive – which it also calls the "Kyoto effect" – the study observes a surge in the patenting of clean energy technologies since the early 2000s. While these findings are for clean technologies, they are also relevant to the discussion on RETs (see box 4.4 for a summary of the key findings).

This increasing tendency to patent is confirmed by other independent analyses of patent trends relating to climate change mitigation technologies. For instance, a recent study found that between 1988 and 2007, Japan had the highest number of claimed priorities for patents in all kinds of climate change mitigating technologies considered in the analysis (Haščič et al., 2010). Japan was followed by the United

Box 4.4: Patents in clean energy: Findings of the UNEP, EPO, ICTSD study

A study by UNEP, the European Patent office and the International Centre for Trade and Sustainable Development (2010) reviews recent evidence on IPRs and RET transfer and ownership, and RET patents for eight clean energy technologies: solar PV, solar thermal, wind, geothermal, hydro/marine, biofuels, CCS and integrated gasification combined cycle (IGCC) clean coal technology. Most of these are RETs, although CCS is used with fossil fuels (coal to be precise).

The review finds that current evidence on whether IPRs constitute a barrier to transfer of RETs is inconclusive. The study also finds that there is relatively little out-licensing of RET patents to developing countries, but that the level is no lower than for other sectors. In addition, it notes that there are the normal sorts of constraints involved in licensing out RET patents, such as those related to high transaction costs, identifying suitable partners and mutual agreement of licensing conditions. Given the urgency to achieve wider diffusion of RETs, the study argues that greater attention should be paid to developing mechanisms that facilitate their licensing. This could be done by improving information flows, reducing transaction costs, expanding capacity in developing countries to negotiate technology licensing and otherwise supporting technology licensing.

Regarding ownership of RET patents, the report finds that patenting of RETs is dominated by OECD countries, although several developing countries also hold some patents, albeit a small proportion. The degree of concentration of patents in the eight technologies considered is found to be similar to that in other technologies, with six OECD countries – France, Germany, Japan, the Republic of Korea, the United Kingdom and the United States – accounting for almost 80 per cent of all patent applications considered in the study. China and Taiwan Province of China are the largest RET patent holders among the non-OECD countries, with most other developing countries accounting for little or none of the total patenting share. Further, the study notes that China and India, despite their growing role in the manufacturing of RETs-related machinery and equipment hold a fraction of the patents held by their Western counterparts.

These findings illustrate that only a relatively small number of countries (or firms in those countries) are generating new patentable knowledge and pushing the technological frontier in RETs. This trend in RETs mirrors a similar trend in patents for technologies more broadly, with few developing countries at the technology frontier of innovation. The developing countries with patents in RETs are relatively advanced, with strong science, technology and innovation capabilities and reasonably well-functioning innovation systems in RETs. No LDCs have any RET patents, if the data reported are accurate.

Source: UNCTAD.

Figure 4.3: Number of patent applications for five renewable energy sources, 1979–2003

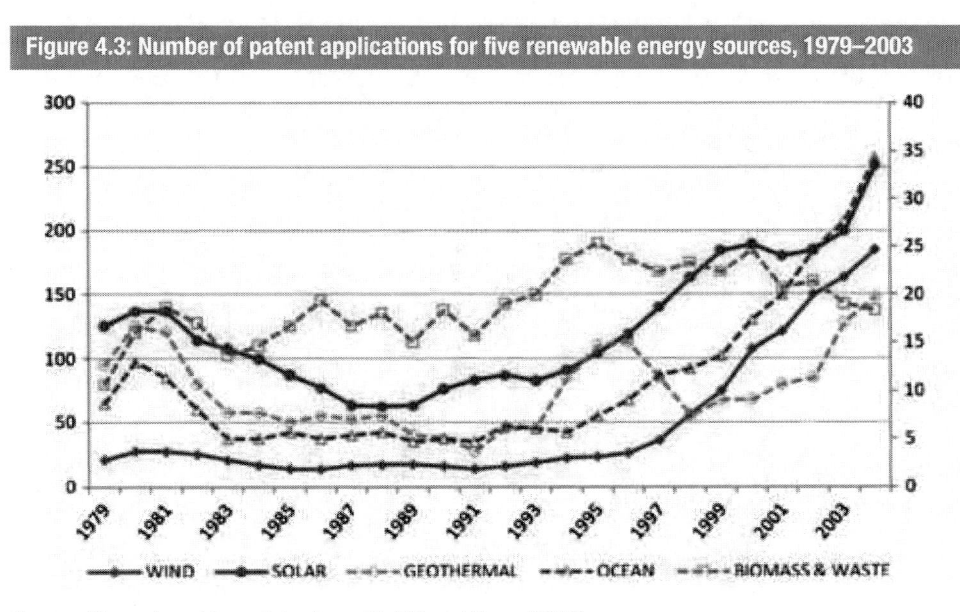

Source: Reproduced from Johnstone, Haščič and Popp (2010).

States, Germany, the Republic of Korea and France. Some smaller countries also figured in particular fields: Denmark for wind technologies, Finland for IGCC clean coal technology and Israel for geothermal technologies. Regarding RE sources, the data show that total patent applications for solar, wind and biomass have been on the rise, especially since 1995 (figure 4.3).

The EU has the highest number of patent applications on RETs (when aggregated for all 27 member countries). Amongst individual countries, the United States accounts

Figure 4.4: Total number of patent applications for energy-generating technologies using renewable and non-fossil sources, 1999–2008

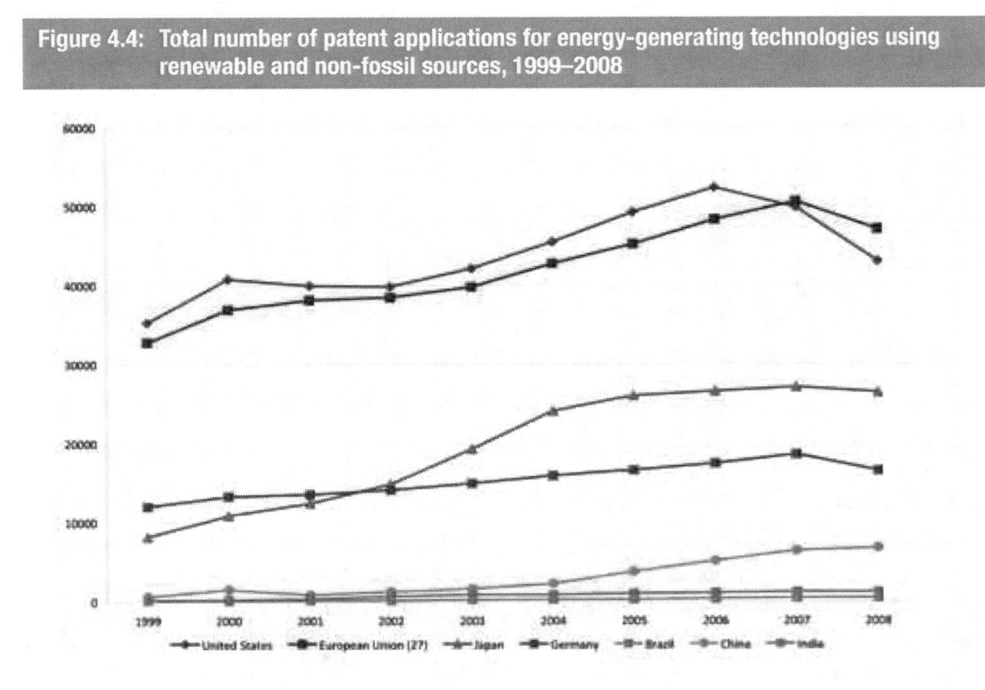

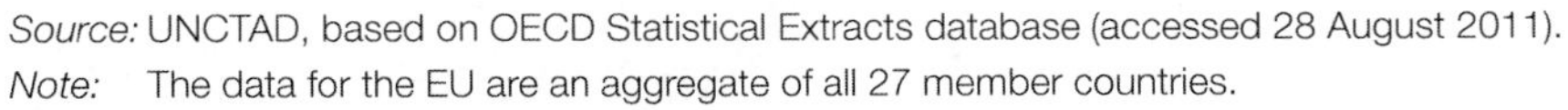

Source: UNCTAD, based on OECD Statistical Extracts database (accessed 28 August 2011).

Note: The data for the EU are an aggregate of all 27 member countries.

[a] Filed under the Patent Cooperation Treaty (PCT).

Figure 4.5: Market shares held by the first, top five and top ten patent applicants in select technologies

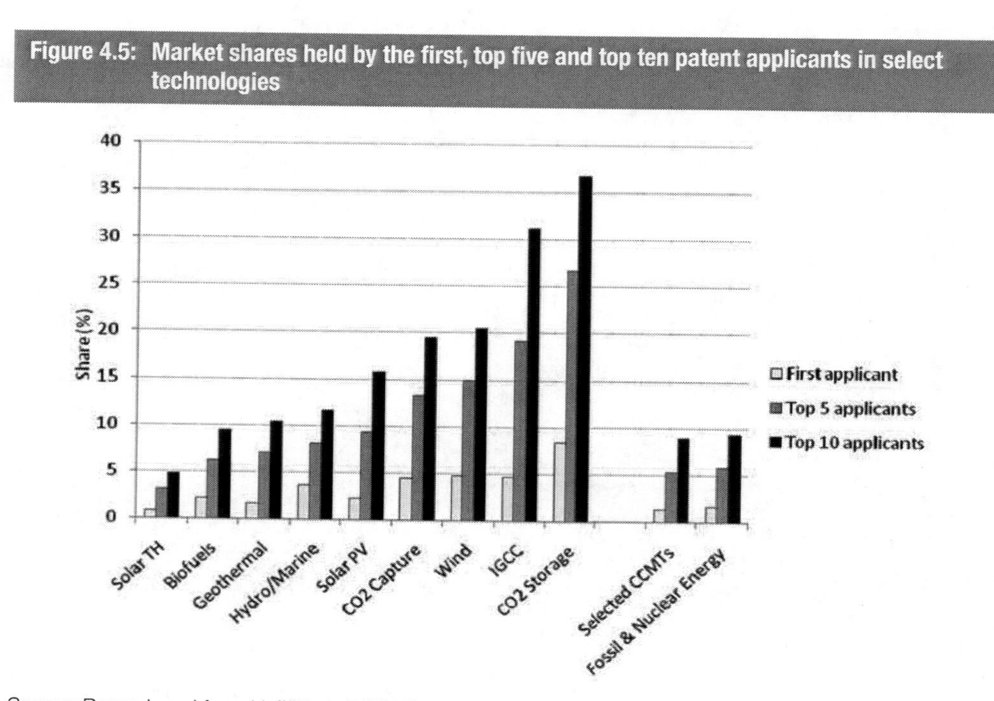

Source: Reproduced from Haščič et al (2010).
Note: CCMTs stands for Climate Change Mitigation Technologies.

Figure 4.6: Countries' shares of patents in solar thermal and solar photovoltaics, 1988–2007 (per cent)

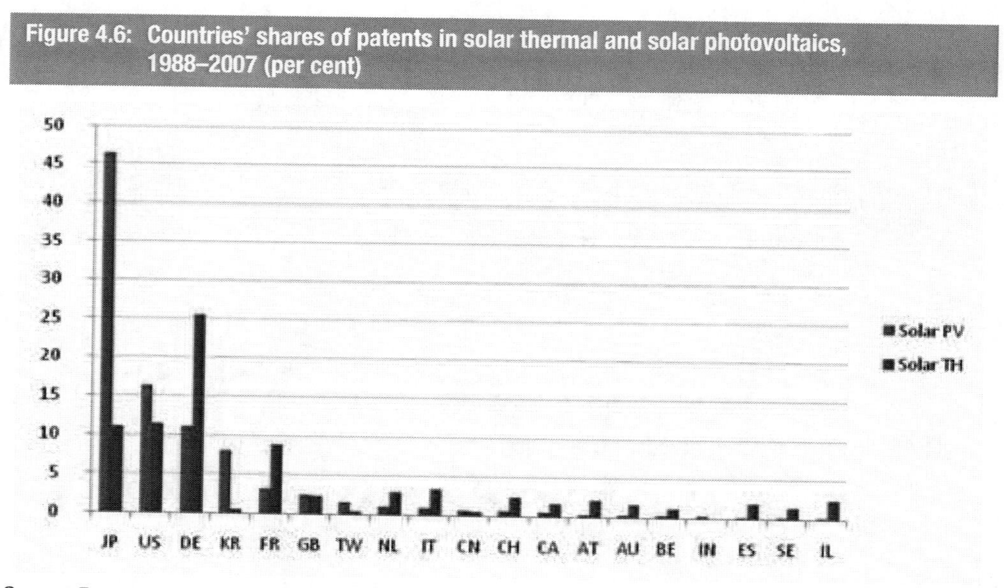

Source: Reproduced from Haščič et al (2010).

Developing countries' shares of patents in solar technologies remain low, despite their growing technological abilities.

for the highest number being filed, followed by Japan and Germany. China, India and Brazil have been increasing the number of patents filed over the past few years, but they figure in the lowest quartile of patent applications (figure 4.4).

Patents can have varying impacts on market power for the patent holders depending on the RET in question. Patents on technologies such as those relating to wind and carbon storage, confer substantial market shares to patent owners compared with those on technologies such as solar thermal and solar PV (figure 4.5). In the case of solar PV, for instance, the top 10 patent holders account for only 20 per cent of the market share. And developing countries' shares of patents in solar technologies remain low (figure 4.6), despite their growing technological abilities.

In conclusion, despite the general finding that most clean technologies needed for developing countries and LDCs are off-patent, patenting activity in RETs is on the rise. In certain RETs, such as those related to wind power, patents seem to confer a large share of the market, indicating a positive relationship between patenting and market access, whereas in some other technologies, such as solar PV, this is not the case.[15] In the case of solar PV, for example, the market structure is largely dominated by products that are not patent protected, and patented products still have a market share of about 20 per cent. Although certain developing countries have developed R&D capabilities in some RETs, the share of patents held by developing-country firms is still low compared with those held by developed-country firms. However, it is important to point out that the statistics do not capture certain incremental innovations and adaptations of RETs in the local developing-country context, unless they are patented. Moreover, licensing contracts often contain more than the simple licence for intellectual property, and licensing contracts are not always registered with government authorities. Also patent mapping does not capture licensing activity and trends, as not all registered patents are licensed. It would be impossible to comprehensively map trade secrets or utility models, as the former are not registered, and, with respect to the latter, many countries do not confer utility model protection over incremental innovations.

Lastly, from these results and from the preliminary trends in patented RETs (discussed below), it is not clear to what extent IPRs will be an incentive for firms to develop RETs or, more broadly, to develop environmentally sustainable technologies in the future.[16]

b. Preliminary trends in patented RETs

A recent study on patent flows and applications for RET-based inventions in countries other than the inventor's home country for solar photovoltaic, wind and biofuel technologies shows emerging trends (Haščič et al., 2010). The study examines the transfer of those technologies from UNFCCC Annex 1 countries to non-Annex 1 countries during the period 1978–2007 (figures 4.7–4.9). The thickness of the arrows in the figures denotes the magnitude of the patents deposited by inventors of these RETs in different parts of the world.

Discussions on technology transfer of RETs in the climate change framework...

The figures are instructive in pointing to some key trends. First, the number of patents being applied for in different countries by the same innovators for solar PV is three times higher than for wind technologies and biofuels. China remains the major patent application country of all technologies being patented in all three areas. Other countries that figure prominently are Brazil, the Republic of Korea and South Africa. Some countries are major recipients of patent applications for particular RETs. For instance, Morocco is a large recipient of patent applications for wind power and Indonesia is an important recipient for carbon capture technologies (not shown in the figures). A second significant trend evident from these figures is that even in 2007 hardly any patent applications flowed from Annex I countries to almost all of northern and sub-Saharan Africa (with the exception of South Africa), large parts of Latin America and South Asia. These trends are important indicators of where and how patent applications are being filed worldwide, reflecting where the learning implications of patent information, if any, could be expected.

...should move beyond focusing narrowly on the transfer of technology to a broader focus on enabling technology assimilation of RETs.

3. Outstanding issues in the debate on intellectual property and technology transfer

The limited data and the ongoing debate on the transfer of environmentally sustainable technologies, and RETs in particular, raises a number of issues relating to technology assimilation, the quality – rather than the quantity – of technology transfer, and flexibilities and other options available to developing countries. These are discussed below.

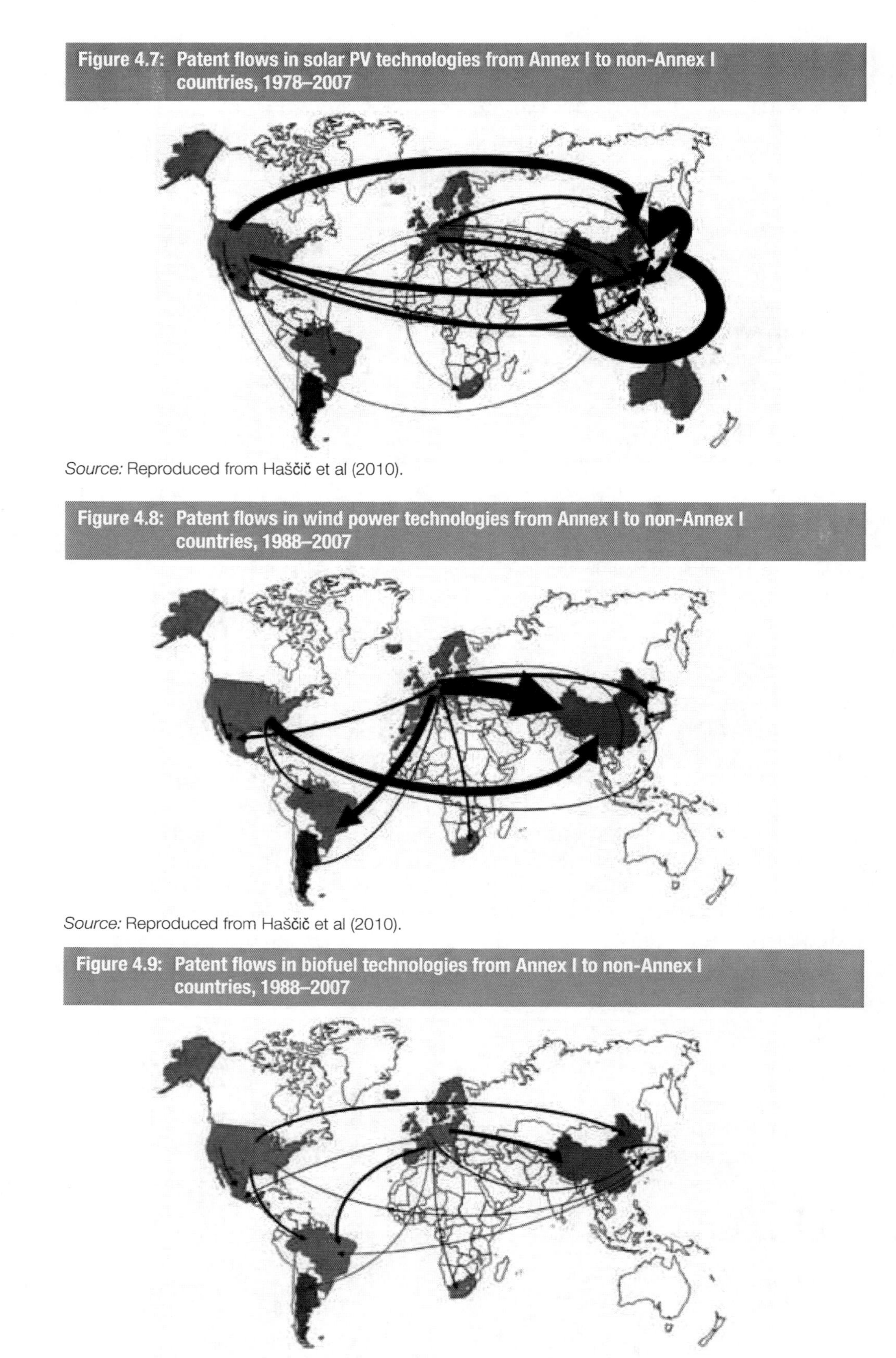

Figure 4.7: Patent flows in solar PV technologies from Annex I to non-Annex I countries, 1978–2007

Source: Reproduced from Haščič et al (2010).

Figure 4.8: Patent flows in wind power technologies from Annex I to non-Annex I countries, 1988–2007

Source: Reproduced from Haščič et al (2010).

Figure 4.9: Patent flows in biofuel technologies from Annex I to non-Annex I countries, 1988–2007

Source: Reproduced from Haščič et al (2010).

a. Beyond technology transfer to technology assimilation

Discussions on technology transfer of RETs in the climate change framework and beyond should move beyond focusing narrowly on the transfer of technology to a broader focus on enabling technology assimilation of RETs. Accumulation of technological know-how and learning capabilities is not an automatic process. Learning accompanies the acquisition of production and industrial equipment, including learning how to use and adapt it to local conditions. In order to promote broader technology assimilation, the technology transfer exercise will need to take into account the specific technological dimensions of RETs as well as the nature of actors and organizations in developing countries. For example, it has been suggested that developing-country research institutions need to be better integrated with development initiatives on RETs, thereby building their capabilities for developing and deploying these technologies, as mentioned earlier in this Report. There is no doubt that many local research institutions in developing countries will have much to offer by collaborating with developed-country research institutions, including undertaking valuable fieldwork to test technologies and adapt them to the local context.

As noted in chapter III, many countries have public research institutions that are actively engaged in RETs research. Regarding the private sector, while small and medium-sized enterprises (SMEs) in developed countries tend to be at the forefront of risk-taking and innovation, this is not the case of similarly placed firms in developing countries. More and more evidence from successful countries and sectors reveals that larger firms innovate much more and are more resilient, and therefore, while SMEs remain very important in developing countries and LDCs, greater support aimed at the expansion of these firms is required. Making technologies more widely available and in ways that promote the national aspirations of countries for technology-led development will be important for expanding the technological base of RETs in developing countries.

b. Assessing the quality – and not the quantity – of technology transfer

The quality of technology transfer should be assessed by the extent to which the recipient's know-how of a product, process or routine activity is enhanced, and not by the number of technology transfer projects undertaken. The impact of any technology transfer initiative in terms of increasing the recipient's know-how varies widely. It depends on the nature of the countries and firms in question (which determines the scope for complementary exchange and learning), the activities under consideration (e.g. prototyping or design) and the technologies involved, as well as on the kinds of activities undertaken as part of the initiative (including training or other forms of capacity building). Technological content in any field is relative, varying according to the domain and the level of the technology that currently defines the frontier. Inventors regularly face the question of whether to disclose an invention in the hope of obtaining a patent, or to keep it protected through other means. For technologies that are developed without the intent to receive any IPRs protection (such as those developed under a grant which prohibits taking out personal IPRs on an invention), there is no database similar to patent searches that would explore and identify existing technologies in the public domain. Even where patent information is disclosed, or even when there are no patents on RETs, it is frequently difficult to find available technologies and to seek transfer or engage in technological learning activities based on such RETs in the absence of domestic capabilities. Finally, often the disclosed information contained in a patent application is insufficient to fully exploit the patent. This is as true for RETs as for other technologies.

The quality of technology transfer should be assessed by the extent to which the recipient's know-how of a product, process or routine activity is enhanced...

...and not by the number of technology transfer projects undertaken.

The nature of licensing possibilities therefore assumes importance for firms that are able to negotiate voluntary licences for technology transfer. In this regard, licensing

The green economy concept...

... gives credence to the view that economic activities need to be environmentally sustainable.

agreements can contain more than just the licence of an IPR, depending on the bargaining position and awareness of the parties involved. They often contain provisions for the transfer of the related know-how required to exploit the invention (which can also be a trade secret). This can take the form of training or specification manuals, for instance.

Negotiation for a technology transfer licence can be a formidable challenge for many developing-country firms. The joint study by UNEP, EPO and ICTSD (2010) on patents and clean energy points out that 70 per cent of respondents to a survey conducted for the study indicated that "they were prepared to offer more flexible terms when licensing to developing countries with limited financial capacity." However, it is unclear whether this would be enough to spur a rapid uptake of RETs in the near future. While RETs markets tend to be relatively competitive, with the possibility of choosing among a range of alternatives, licensing royalties and other restrictive terms commanded by owners of IP could still be problematic for many developing countries.

c. Exploring flexibilities and other options within and outside the TRIPS framework

At the international level, including at the UNFCCC intergovernmental discussions, it has been suggested that flexibilities similar to those advocated for promoting access to public health and medicines could be used in order to enable greater access and diffusion of RETs in developing countries (UNCTAD, 2009).[17] To some extent, such discussions have focused on the use of compulsory licences as one means of supporting greater and more rapid transfer of RETs. A number of recent studies point out that there may be important differences between pharmaceutical technologies and RETs. For instance, in the pharmaceutical sector, an individual patent may have a very substantial impact because a specific drug may not have substitutes (Barton, 2007) and creating a substitute may require years of uncertain R&D investments. Such a situation grants the patent holder a very strong market position, and a compulsory licence is often a powerful negotiating tool that may help in obtaining favourable terms for technology transfer licences or cheaper drug prices. A second difference is that in the case of pharmaceuticals, non-availability or non-affordability of a particular drug in developing countries due to patent protection can result in direct losses of health and life.

D. THE GREEN ECONOMY AND THE RIO+20 FRAMEWORK

For much of the industrialized world, issues related to climate change have begun to revolve around the notion of the "green economy". Still an evolving concept, the green economy can be defined as economic development that is cognizant of environmental and equity considerations while contributing to poverty alleviation.[18] A report by a Panel of Experts to the United Nations Conference on Sustainable Development notes that the concept has gained currency in the light of the recent multiple crises in the world (climate, food and financial), as a means of promoting economic development in ways that "...will entail moving away from the system that allowed, and at times generated, these crises to a system that proactively addresses and prevents them" (UN/DESA, UNEP and UNCTAD, 2010: 3).

The green economy concept is not entirely novel; in many ways, it builds on the well-known notion of sustainable development, and has gained ground as a multilateral negotiating process within the Rio+20 framework. Fundamentally, it gives credence to the view that economic activities need to be environmentally sustainable, and that therefore it is imperative to factor in all environmental externalities of modern day processes. Most industrialized countries view this in terms of regulating economic activities (individual or firm-level) to reduce their carbon emissions. However, it is important to make sure that such regulatory measures do not end up blocking imports of goods and services from developing countries on the grounds of their non-compatibility with the green economy.

How far this can actually be made to happen in an inclusive way is still much debated. For most developing countries, the issue still remains one of how constant technological learning and innovation that underlies structural transformation can be ensured using the possibilities offered by RETs (as analysed in chapter III). In the specific context of this report, what remains important is that the emerging standards on carbon footprinting and border carbon adjustments are not used in ways that may be inimical to developing countries.

1. Emerging standards: Carbon footprints and border carbon adjustments

Border carbon adjustments involve trying to level the playing field between domestic producers that are subject to carbon pricing and foreign producers that face lower regulatory costs. Carbon footprinting standards are country-based standards that seek to reduce the carbon footprint of economic activities by setting standards on the amounts of GHG emissions caused by an individual, organization, firm or product.

Product carbon footprint labelling involves labelling of goods, either on a voluntary or mandatory basis, to inform consumers of the amount of carbon that is embodied in their life cycle. Increasingly, private sector labels are being used by major retail

It remains important that the emerging standards on carbon footprinting and border carbon adjustments are not used in ways that may be inimical to developing countries.

Table 4.3: Companies using the United Kingdom's Carbon Reduction Label (by April 2011)

Company name	Product description
Dyson	Hot air hand dryer
Walkers	All varieties of standard crisps sold in single packets
Tate and Lyle	1kg bag of granulated cane sugar
Tesco	Range of toilet paper and kitchen roll
Tesco	Milk: skimmed, semi-skimmed, whole
Tesco	Range of own-brand laundry detergent
Tesco	Range of chilled and long-life orange juices
Tesco	Range of light bulbs
Tesco/MMUK	Jaffa oranges / soft fruit
PepsiCo	Quaker oats and Oats-so-Simple
Morphy Richards	Range of irons
Allied Bakeries	Kingsmill wholemeal, white and 50:50 loaves of bread
British Sugar	A range of white granulated sugar – British Sugar - B2B
British Sugar	A range of white granulated sugar – Silver Spoon - B2C
Haymarket	Magazines – Marketing and ENDS report
Continental Clothing	A range of over 800 t-shirts and other cotton apparel
Continental Clothing	Woven bags (United States and Japan) and t-shirt Internet retailing service
Marshalls	Complete range of 2,500 paving products
Mey Selections	Scottish honey and shortbread
Sentinel	Central heating cleaning fluid
Stalkmarket	Biodegradable, disposable catering service packaging
Aggregate Industries	Three varieties of paving products – Bradstones
Axion	Recycled plastic/polymer
Baxter	Flexbumin (health care)
Suzano	Pulp and paper products

Source: UNCTAD, based on Cosbey and Savage (2011).

chains in some OECD countries, but there are several methodological problems with these, primarily concerning their scope. For instance, it is unclear whether their calculations take into account the emissions from land conversion to crops (Plassmann et al., 2010), or whether they focus only on transport (so-called "food-miles" labels) – a criterion that inherently discriminates against foreign producers. Private sector voluntary standards are on the rise, gradually becoming an important factor for developing-country exporters to gain markets in some sectors (Potts, van der Meer and Daitchman, 2010). Current schemes focus primarily on agriculture and agri-food products, but increasingly a number of consumer manufactures are being included. The United Kingdom's Carbon Reduction Label, managed by the Carbon Trust, covers a wide range of products (table 4.3). Another problem with the labeling currently is that they use different methods to calculate the carbon content, pointing to the need for clear benchmarks using which such schemes can be devised.

The global move towards climate change mitigation needs to be accompanied by international finance and technology support measures...

In June 2010, France passed an environmental bill (Grenelle 2) and one of its articles mandates carbon footprint labelling for a range of products. Details of the methodologies to be used have yet to be determined. Similarly, the EU has initiated studies on carbon footprint labelling to assess possible methodologies and features that might be "relevant for future policy development" (Swedish National Board of Trade, 2010). It is likely that these studies will eventually lead to some sort of EU-wide labelling scheme.

...that help developing countries and LDCs to transition to RETs in a strategic and sustainable manner.

Similarly, border carbon adjustment (BCA) has been proposed by a number of analysts, and has been incorporated in one form or another in every piece of United States climate legislation to date.[19] Some EU countries are actively promoting it in the context of the third phase of the EU's Emissions Trading Scheme. It involves assessing a tax adjustment at the border, or requiring importers to buy into a domestic cap-and-trade regime for carbon emissions. The legal status of such regimes is not possible to assess *ex ante*, as much depends on the details of the specific policy, and there is no legal consensus on the effectiveness of BCAs as an policy tool within the WTO (Cosbey, 2009).

Both carbon labelling and border carbon adjustments operate by trying to assess the embodied carbon in goods at the border, and sanctioning accordingly. To the extent that labelling continues to spread, and BCAs are implemented by major importing countries (neither of which is certain), developing countries will be forced to resort to the use of RETs to comply with import restrictions. However, simply forcing developing countries to use RETs through measures such as carbon labelling and BCAs may not be sufficient to enable the transition. Indeed it may even have adverse effects on industries in developing countries by acting as barriers to imports, since enterprises and organizations may not have the means (financial and technological) to meet these new requirements. To ensure that industries in developing countries and LDCs are not additionally saddled with "green" requirements, the global move towards climate change mitigation needs to be accompanied by international finance and technology support measures that help developing countries and LDCs to transition to RETs in a strategic and sustainable manner. This is very important to ensure that the green economy concept does not impose additional constraints on the industrial performance of these countries.

2. Preventing misuse of the "green economy" concept

In addition to providing critical infrastructure to support the emergence and shift in production structures within developing countries, RETs can serve the goals of industrial policy by helping those countries' exporters become more competitive in the face of increasingly stringent international environmental standards. To ensure a smooth and equitable transition of all countries to RET-based energy use, it is important to identify and cope with the trade-offs posed by the green economy concept for developing countries and LDCs. The principle of common-but-differentiated responsibilities, which has become a significant cornerstone of the climate change negotiations, provides the best starting point for this.

Developing countries and policymakers have raised concerns about possible misuse of the green economy concept, particularly since adoption of this concept without considering its negative effects on developing countries could worsen their situation. Developing countries may become vulnerable to protectionist policies that have environmental objectives (UN/DESA, UNEP and UNCTAD, 2010). It has been proposed that the stage of development of countries and their ability to cope with the requirement to shift to "green" production and consumption processes need to be considered when implementing this new development paradigm.

E. FRAMING KEY ISSUES FROM A CLIMATE CHANGE-ENERGY POVERTY PERSPECTIVE

As mentioned earlier in this chapter, there is an urgent need for a shift in positioning issues within the international agenda. The obligations of countries to mitigate climate change should be framed in terms of environmentally sound development opportunities for all. Such a repositioning implies actively supporting RETs in developing countries through financial, technology transfer and technology acquisition measures. This section presents options for consideration within ongoing discussions on financing climate change adaptation, but they may also be considered in separate and new initiatives that focus particular attention on the needs of developing countries.

1. Supporting innovation and enabling technological leapfrogging

A limited number of developing countries are steadily making a mark as developers of RETs and their firms are gaining significant markets in RETs, as discussed in chapter III. Some studies have also noted that expertise in developing countries has been concentrated largely in the lower tech RETs such as biofuels, solar thermal and geothermal, in which they have either existing expertise, or good chances of developing competitive exports (Steenblik, 2006). Furthermore, in China and India, the sizeable domestic markets have been springboards for their subsequent success in exports, driven, as in the OECD countries, by ambitious domestic targets for renewable energy generation. China installed 16.5 GW of domestic wind power capacity in 2010 – more than any other country, and more than three times the amount installed in the United States (Ernst & Young, 2011). India ranked third with a capacity addition of 2.1 GW (Balachandar, 2011).

Advantages of using RETs will not accrue automatically.

Still, most RETs are developed and produced by industrialized countries (IPCC, 2007). As a result, firms in developing countries, which are largely technology followers in this field, tend to underinvest, or they have difficulties in accessing technologies and related know-how from abroad and in learning how to use it effectively. Most proponents of the leapfrogging argument (e.g. Gallagher, 2006) tend to posit that since technologies are already available, they can be used at marginal costs by developing countries and LDCs to circumvent being "locked into" the conventional, resource-intensive patterns of energy development. It is also claimed that leapfrogging is possible because RETS can contribute to building new long-term infrastructure, such as for transport and buildings, in ways that promote cogeneration of technologies (Holm, 2005).[20] These are indeed advantages of using RETs, but they will not accrue automatically. Tapping into already developed technologies and related knowledge that

There is a greater need to make the international trade and IPR regimes more supportive of the technological requirements of developing countries and LDCs.

already exists call for capabilities for dynamic learning and innovation within countries. Therefore, promoting greater access to RETs, along with increased support for their use and adaptation through all means possible, will be important for developing countries to sustainably integrate these processes alongside efforts aimed at capital formation and structural transformation.

Apart from the need for strong domestic technology and innovation policies for RETs, there is also a greater need to make the international trade and IPR regimes more supportive of the technological requirements of developing countries and LDCs.

Four mechanisms of support could be made available at the international level.

The obvious question for all developing countries and for the global community is whether the BRICS countries (Brazil, the Russian Federation, India, China and South Africa) are special cases. To some extent they are: they have the prerequisites for competitive production of many RETs, such as a workforce with advanced technical training, supporting industries and services in high-tech areas, access to finance, ample government assistance and a large domestic market, all of which would seem to favour these larger emerging developing countries over smaller, poorer developing countries and LDCs. Historically, promoting technological learning and innovation has remained a challenge for all developing countries. The experiences of China, India and other emerging economies show that public support, political will and concerted policy coordination are key to promoting technological capabilities over time. Greater support for education (especially at the tertiary level), and for the development of SMEs, as well as financial support for larger firms and public science are particularly important. While these factors have to do with domestic policy, greater support from the international community is also imperative.

Enhancing RETs-based learning in LDCs with the purpose of promoting energy access would fit in directly with the International Innovation Network agreed at UN LDC IV.

This *TIR* proposes four mechanisms of support that could be made available at the international level, namely: an international innovation network for LDCs, with a RETs focus, global and regional research funds for RETs deployment and demonstration, an international technology transfer fund for RETs and an international training platform for RETs. The international innovation network for LDCs has already been agreed upon in the LDC IV Conference held in Istanbul in May 2010. This *TIR* suggests a RETs specific focus for this endeavour. The other three mechanisms could be considered by the international community both within and outside of the climate change framework.

a. An international innovation network for LDCs, with a RETs focus

Existing initiatives and mechanisms available to support the accumulation of technological know-how and technology transfer of RETs are neither comprehensive nor specifically tailored to the particular needs of developing countries and LDCs. Some of them only provide information on available technologies without supporting technology transfer. However, as pointed out earlier in this chapter, the acquisition of information on the technologies available is only one part of the process of transfer. To address some of the shortcomings of existing initiatives on technology transfer, a science, technology and innovation centre (an International Innovation Network) was proposed at the UN LDC IV Conference in Istanbul in May 2011.

The Istanbul Declaration states that the proposed International Innovation Network is intended to "...promote access of LDCs to improving their scientific and innovative capacity needed for their structural transformation." The Centre, a signature initiative of the UN LDC IV Conference, is intended to serve as a real and virtual hub for, among others: "Facilitating joint learning – through exchange of information and experiences as well as establishment of a shared knowledge base of analytically rigorous, shared case studies – to enable peer-to-peer learning between experts, organizations and agencies from LDCs and other countries with recent and ongoing development experiences."

The knowledge-sharing activities of the Network are intended to focus on four key areas, one of which would promote technological leapfrogging for facilitating access to energy by building combined clean energy and ICT-networked infrastructures. This *TIR* suggests that a specific focus on enhancing RETs-based learning in LDCs with the purpose of promoting energy access would fit in directly with the currently identified key areas.

Such a RETs-based area of work, as this Report suggests, could focus on:

(i) Promoting a network-based model of learning and sharing of experiences of countries on how to increase energy access through RETs in rural, mini-grid areas;

(ii) Promoting access to financing opportunities and technology licences needed for upgrading the RE technological base in the private sector in LDCs, by establishing partnerships with both international firms and donor agencies; and lastly,

(iii) Establishing an information-sharing mechanism where different kinds of stakeholders could network and work together to enhance the knowledge base in LDCs on RETs use, adaptation, production and innovation.

b. Global and regional research funds for RETs deployment and demonstration

RETs are a key area of particular interest to developing countries that has lacked funding for technology development and demonstration. As chapter III has noted, public expenditures in this area in even the more advanced developing countries has been relatively stagnant. The dedicated funds, in their designated organizational structures, could act as the focal point for the coordination of ongoing research both at the national and regional levels, and among private, public and non-profit organizations. The funds could also ensure open access to all available research.

Scaled up technical cooperation and training programmes could complement these funds, involving skilled workers from both developed and emerging economies (engineers, teachers and technicians, among others) moving on a temporary basis to help develop local capacity in developing countries and LDCs. These kinds of initiatives are already under way in other areas such as public health.[21]

Dedicated funds for RETs deployment and demonstration could act as the focal point for coordination of ongoing research both at the national and regional levels.

Regional R&D facilities which focus on technological improvements aimed at making RETs cost-effective would be an important part of such initiatives. Such facilities could be created by developing countries and LDCs themselves or through South-South collaboration, or even as a triangular facility between developing countries and LDCs (offering and receiving technical know-how and training) and developed countries (offering financial assistance). Such facilities have been a core component of industrial sector policies in several economies, including China, India, the Republic of Korea and Taiwan Province of China. The regional R&D funds could also set research priorities for technological expansion of RETs by firms.

c. An international technology transfer fund for RETs

The limited data analysed earlier in this chapter shows a trend of proliferation of patents in RETs which may lead to a more skewed and unfair distribution of future opportunities for firms in developing and least developed countries in these technological domains. Not only do firms in developing countries and LDCs find it difficult to search and inform themselves of appropriate RETs for technological acquisition, they may often lack the capacities required to negotiate licences for the technologies in question. Bargaining costs of acquiring licences can be extremely high. Firms also lack information on the kinds of similar technologies available, and their relative costs and merits, which limits their ability to make informed choices.

A technology transfer fund for RETs could address issues by acting as a licensing pool for technologies.

A technology transfer fund for RETs could address all three of these issues by acting as a licensing pool for technologies, which would offer the RETs at a subsidized rate for firms from LDCs and from developing countries with low technological capabilities. Funds generated by such developing countries and LDC governments themselves, or by donor agencies, or both jointly, could be used to subsidize the licensing costs to the recipient firms. The fund could also provide a database of similar technologies and their relative merits and licensing costs, thereby creating a much-needed service for firms and organizations in developing countries and LDCs. By acting as a clearing house for the licensed technologies, it would also reduce bargaining asymmetries between firms in developed countries on the one hand, and those in developing countries and LDCs on the other.

As incentives for firms in the industrialized countries to participate in the RET technology transfer fund, the initiative could provide the market fees of licensing for the firm, in addition to guaranteeing internationally agreed standards of IPR protection. The firms in industrialized countries that participate in the fund could also receive a label (similar to eco-labelling) indicating that they support "pro-green economy", thereby procuring goodwill from global markets, similar to "fair trade" labels. Firms from developing countries and LDCs that qualify to receive the technologies transferred would be subsidized according to their ability to pay. To this end, the initiative could set a series of financial thresholds that would determine the amount that the recipient firms will be charged for the technologies available as part of the fund.

Establishing an international training platform specifically for RETs... ...would serve the important goal of creating a skilled staff base across developing countries for their wider use and promotion.

The RETs technology transfer fund would be different from patent pooling in two important respects. The fund would provide licences not only for patented products, but also for products that are protected through other forms of IP, thereby covering a wide range of sectors and firms. Second, the fund would not rely on the altruistic motives of firms in industrialized countries; the firms that own the technology transfer licences would stand to gain from "pro-green economy" labelling in addition to receiving the market price for the licences.

d. An international training platform for RETs

Use and adoption of RETs is a relatively new process in many developing countries and LDCs. The underlying processes for this could be significantly facilitated through appropriate training and skills creation, which are presently under-provided in many developing countries and LDCs. As noted in chapter III, the availability of skilled personnel for installation and maintenance of RETs is important for making them cost-effective. Further, the availability of trained staff in various aspects of RETs use would promote general awareness of the technologies and enable countries to exploit their versatility for combination with conventional energy sources, while also assuring their proper maintenance and optimal use.

Establishing an international training platform specifically for RETs, as proposed by this *TIR,* would serve the important goal of creating a skilled staff base across developing countries for the wider use and promotion of RETs in domestic and industrial contexts. The proposed international RETs training platform could operate at two different levels: a physical institute based in one or several places throughout the world, which would offer training on various aspects of RETs use, adaptation and production; and a virtual training platform offering online, computerized courses of various kinds. In both forms of training (through the physical centers and the virtual training platform), RETs-related learning could involve various fields, such as material sciences, marketing, legal issues, energy combinations and RET applications in various industrial fields.

2. Coordinating international support for alleviating energy poverty and mitigating climate change

International cooperation is important for various aspects of renewable energy, such as for ensuring the availability of comparable and comprehensive statistical data sets to inform policy, promoting technology transfer and innovation, capacity-building and financing. So far, international support has been fragmented, with various international organizations working in the field, including UNEP, the United Nations Industrial Development Organization (UNIDO), UNCTAD and several other agencies active in particular aspects of RETs use and promotion. While some have emphasized the climate change dimension, others have stressed the green economy and yet others have focused on RETs.

A dedicated international agency, the International Renewable Energy Agency (IRENA), was established in 2010 with the specific purpose of promoting the widespread use and adaptation of RETs, as well as for dealing with issues of renewable energy-related innovation by promoting greater dialogue (box 4.5). IRENA could also serve to coordinate international support, which could be an important step in promoting joint efforts in this area.

3. Exploring the potential for South-South collaboration

The success of many developing countries in the development and production of RETs implies the existence of considerable potential for technology flows, trade, investment and cooperation among these countries. They provide a huge market for such technologies and capturing even a portion of it would bring significant development benefits. Thus, intra-South trade in RETs and/or investment in the production of RETs could contribute towards building globally viable domestic capacity of developing countries in the sector. Given the existing patterns of intraregional trade, it may

Box 4.5: The International Renewable Energy Agency

The statute creating IRENA was adopted on 26 January 2009, and entered into force on 8 July 2010. IRENA's purpose is to promote the widespread and increased adoption and sustainable use of all forms of RE (biomass, geothermal, hydro, marine, solar and wind). The statute of IRENA has been ratified by 76 States (of its 148 members) and the EU. It took only 27 months from the founding conference to the creation of the Agency, which reflects the high degree of importance accorded to this area by various countries. IRENA is the only agency with a global mandate for RE.

The first session of the IRENA Assembly, held on 4–5 April 2011 in Abu Dhabi, mandated the Agency to play a strong role, regionally and globally, to support countries in accelerating their adoption of renewable energy as a key component of national, subregional and regional development plans.

IRENA's Work Programme for 2011 includes action on three key fronts: the Knowledge Management and Technology Cooperation (KMTC) sub-programme is designated to facilitate an increased role for RE; the Policy Advisory Services and Capacity Building sub-programme seeks to stimulate an enabling environment for uptake of RE; and the Innovation and Technology sub-programme aims at creating a framework for technology support, identification of the potential for cost reduction and the wider use of standards.

In order to assist governments in their efforts to develop efficient and effective RETs and innovation strategies, the IRENA Innovation and Technology Centre (IITC) is collecting and analysing data on scenarios and strategies with the aim of transforming them into policy-relevant information for decision-makers. Additionally, this information will be used by the renewables readiness assessments expected to be undertaken as part of the Knowledge Management and Technology Cooperation sub-programme. At present, the main focus is on Africa, but preparations have begun for a similar undertaking in the Pacific region. One of the essential elements for greater deployment of REs in developing countries is technology transfer and dissemination. This could be facilitated by greater use of the existing patents that could be made available to these countries at little or no cost. A website on patent search is under development, which will include tools for simplifying the search functions and utilities.

IITC has commenced its work on technology road mapping with the aim of identifying prospects, technological barriers, financing, and development and policy needs. In order to gain a better understanding of the costs involved in the wider use of RETs across countries, and how technology development could potentially help reduce those costs, the IITC is expected to report on aspects of power generation using RETs. This information is expected to assist the member countries in their investment decisions and in identifying opportunities for further cost reductions. Further expansion of such a reporting system to other end-use sectors of the economy (such as transport) is planned for the future.

Source: IRENA for *Technology and Innovation Report 2011*.

also be an easier starting point for developing countries' exports to, or investments in, developed countries. However, there are some obstacles to the export of RETs from developing countries, including inadequate marketing channels and expertise. These challenges are particularly acute for SMEs and new entrants (Semine, 2010). However, increased trends in South-South cooperation imply that such problems may not be too daunting for many developing-country firms.

Another major benefit of South-South trade and investment is that developing-country partners have similar needs for many technologies, at least more so compared with developed-country markets. Solar cooking stoves, for example, are an innovation well suited to developing-country contexts, where there is generally ample sunshine, and they are uniquely adapted to the needs of developing-country consumers, many of whom have been forced to rely on traditional biomass for cooking. The same may be said for solar PV-powered lights and lanterns, the biggest markets for which are poor rural locations without grid access. As such, developing-country partners can provide a market for RETs that have been developed to serve domestic needs or for non-indigenous technologies that have been adapted to serve local needs. Of course this is not equally valid for all RETs; installation of wind turbines, for example, does not vary among developed and developing countries.

A major benefit of South-South trade and investment is that developing-country partners have similar needs for many technologies.

F. SUMMARY

This chapter has analysed the key issues in international policy-making that are important for developing capabilities in all aspects of RETs use and innovation in developing countries and LDCs. Focusing on finance, technology transfer and IPRs, it shows how incoherent and often conflicting policy developments at the international level tend to adversely affect national aspirations for technological empowerment in developing countries. The chapter argues for an international agenda that focuses equally on climate change mitigation and energy poverty alleviation. It makes the case for more concerted international support for RETs use, adaptation and innovation in developing countries, including the following:

(i) Financing for RETs needs to be conceived and implemented both within and outside the climate change framework as a priority. This would serve the dual need of mitigating climate change as well as supporting economic development in developing countries. The positive implications of RETs use for development need to be better included in the discussions on the financing of climate change mitigation efforts. The international financing architecture for climate change mitigation needs to be redefined accordingly, and additional measures for the financing of RETs introduced.

(ii) The technological empowerment of developing countries and LDCs to use, adapt and innovate in RETs should be the fundamental goal of technology transfer in this area. This should be based not on the number of ongoing technology transfer projects at any given point in time, but rather on an assessment of the quality of technologies transferred. A parallel emphasis on strengthening regulatory frameworks, building institutional capacity and enhancing the absorptive capacity of recipient countries is also necessary.

(iii) Greater international support in the area of technology and innovation for RETs could take the form of several important initiatives. The chapter has proposed four such initiatives, namely, an international innovation network for LDCs, with a RET focus, global and regional research funds for RETs deployment and demonstration, an international RETs technology transfer fund and an international RETs training platform. More support could take the form of augmenting and further strengthening the recently proposed technology mechanism within the UNFCCC, so as to strengthen its focus on RETs.

Building further on this analysis, the next chapter proposes incentives as part of an integrated innovation policy framework within countries to achieve these goals at the national level.

NOTES

1 Broadly speaking, the processes that fall *under adaptation* are those that seek to reduce/prevent the adverse impacts of ongoing and future climate change. These include actions, allocation of capital, processes and changes in the formal policy environment, as well as informal structures, including social practices and codes of conduct. *Mitigation* of climate change, on the other hand, seeks to prevent further global warming by reducing the sources of climate change, such as GHG emissions.

2 "Clean technologies" or "clean energies" cover a much broader range than RETs, and include clean coal, for example.

3 For instance, on the one hand, countries are expected to grant IPRs in accordance with the TRIPS Agreement that restrains their access to patent protected technologies, but at the same time, the climate change framework calls for greater access to technologies, whether or not such technologies are patent protected.

4 "Clean technologies" or "clean energies" is generally a much broader concept than RETs, and includes clean coal.

5 The Climate Change Convention is currently in its draft form, awaiting formal ratification and adoption.

6 See Article 4(3), 4(4) and 4(7) of the Draft Climate Change Convention.

7 Ocampo and Vos (2008), for example, note that while some developing countries have achieved significant progress over the past three decades, several others have remained stagnant.

8 The relevant provision reads thus: "In the case of privately owned technologies, the adoption of the following measures, in particular for developing countries: iv. In compliance, with and under the specific circumstances recognized by, the relevant international conventions adhered to by States, the undertaking of measures to prevent the abuse of IPRs, including with respect to their acquisition through compulsory licensing, with the provision of equitable and adequate compensation." UN/DESA (2009b) also makes an implicit reference to technology transfer in the context of the environmental and development goals of the 1972 United Nations Conference on the Human Environment.

9 UNFCCC 2007, Bali Action Plan, Document FCCC/CP/2007/L.7/Rev.1.

10 Prior to the TRIPS Agreement in 1994, not many developing countries and LDCs provided the same standards of protection as were required by the subsequent Agreement. Their patent protection terms were much shorter than the 20 years mandated by the Agreement and national patent laws contained several provisions which are not allowed under the Agreement, such as a "working" requirement that mandated that inventions be produced domestically in order to qualify for grant of a patent.

11 These are provisions that can be used to nuance the impacts of IPRs on domestic regimes for technological learning and industrial development. Several such flexibilities exist in the TRIPS Agreement. For a discussion of the key TRIPS flexibilities and how they can be used, see for example, Reichmann, 1996; Correa, 2000; and CIPR, 2002.

12 See the discussions under the auspices of the Commission on Intellectual Property Rights, Innovation and Health (CIPIH) and negotiations leading to the Global Strategy and Plan of Action of the World Health Organization (WHO), which came into force in 2009 (www.who.int/gb/ebwha/pdf_files/A61/A61_R21-en.pdf)

13 See discussions relating to the International Convention for the Protection of New Varieties of Plants (UPOV Convention, 1991 version), the rights of farmers in developing countries and the sui generis option under Article 27(3)(b) of the TRIPS Agreement.

14 The Cancun negotiations in 2010.

15 The data only show patented products in the wind and solar markets. When these markets are considered in their totality (i.e. including inventions that are off-patent), individual firms may have smaller market shares (denoting the absence of an oligopolistic market structure). Although the data for conducting such an analysis are currently unavailable, this needs to be borne in mind.

16 A recent study concludes that the patenting system does not drive significant levels of R&D in most environmentally sustainable technologies (Maskus, 2010). The study reveals that IPRs alone are insufficient incentives for small technological solutions in uncertain markets.

17 Brazil, China and India have advocated stronger use of TRIPS flexibilities at the UNFCCC intergovernmental meetings, including the greater use of compulsory licences.

18 In the ongoing negotiations relating to the Rio+20 framework, the concept of green economy is viewed in the context of sustainable development and poverty reduction.

19 For an overview of BCAs, their effectiveness and the elements of the proposed regimes, see Wooders, Cosbey and Stephenson, 2009.

20 Cogeneration of technologies refers to the possibility of developing new (but complementary) sets of technologies in parallel, as explained in chapter I.

21 A good example is the Engineering Capacity Building Programme by the Deutsche Gesellschaft für Internationale Zusammenarbeit (GIZ). As part of this program, a bioequivalence facility for the East African Region is being set up in collaboration with two pharmaceutical companies from Kenya, one from Tanzania and one from Ethiopia and the School of Pharmacy, University of Addis-Ababa.

REFERENCES

Arora A and Gambardella A (2004). The globalization of the software industry: Perspectives and opportunities for developed and developing countries. Cambridge, MA, National Bureau of Economic Research, June.

Balachandar G (2011). India closely follows China, US in wind power capacity addition. Available at: http://www.mydigitalfc.com/news/india-closely-follows-china-us-wind-power-capacity-addition-752 (accessed 8 August, 2011).

Barton JH (2007). Intellectual property and access to clean energy technologies in developing countries: An analysis of solar photovoltaic, biofuel and wind technologies. Geneva, ICTSD.

Birdsall N and Subramanian A (2009). Energy needs and efficiency, not emissions: Re-framing the climate change narrative. Washington, DC, Center for Global Development, November.

CIPR (2002). Integrating Intellectual Property Rights and Development Policy, London. Available at: http://www.iprcommission.org/papers/pdfs/final_report/CIPRfullfinal.pdf

Correa CM (2000). *Intellectual Property Rights, the WTO, and Developing Countries: The TRIPS Agreement and Policy Options*. London, Zed Books.

Correa CM (2005). Can the TRIPS Agreement foster technology transfer to developing countries? In: Maskus KE and J. H. Reichman JH, eds. International Public Goods and Transfer of Technology under a Globalized Intellectual Property Regime. New York, Cambridge University Press: 227–256.

Cosbey A (2009). A sustainable development roadmap for the WTO. Manitoba, International Institute for Sustainable Development.

Cosbey A and Savage M (2011). RETs for sustainable development: Designing an integrated response for development. Background paper for *TIR 2011.* Geneva, UNCTAD.

Dechezleprêtre A, Glachant M and Ménière Y (2008). The Clean Development Mechanism and the international diffusion of technologies: An empirical study. *Energy Policy*, 36(4): 1273–1283.

Ernst & Young (2011). Renewable energy country attractiveness indices no. 28.

Gallagher K (2006). Limits to leapfrogging in energy technologies? Evidence from the Chinese automobile industry. *Energy Policy*, 34(4): 383–394.

Global Humanitarian Forum (2009). *Human Impact Report: Climate Change–The Anatomy of a Silent Crisis*. Geneva. Available at: http://www.ghf-ge.org/human-impact-report.pdf.

Haščič I, Johnstone N, Watson F and Kaminker C (2010). Climate policy and technological innovation and transfer. An overview of trends and recent empirical results. OECD *Environment Working Papers*, No. 30, OECD, Paris. Available at: http://www.oecd-ilibrary.org/docserver/download/fulltext/5km33bnggcd0.pdf?expires=1313138057&id=id&accname=guest&checksum=5F9A4AE055F43297CA7AEFFF51EB78BE

Holm D (2005). Renewable energy future for the developing world. ISES White Paper, International Solar Energy Society, Freiburg.

IEA (2000). *Energy Technology Perspectives 2000*. Paris, OECD/IEA.

IEA (2010a). *World Energy Outlook 2010*. Paris, OECD/IEA.

IEA (2010b). *Energy Technology Perspectives 2010.* Paris, OECD/IEA.

IPCC (2007). *Climate Change 2007: Mitigation of Climate Change.* Contribution of Working Group III to the Fourth Assessment Report of the Intergovernmental Panel on Climate Change. Cambridge and New York, Cambridge University Press.

Johnstone N, Haščič I and Popp D (2010). Renewable energy policies and technological innovation: Evidence based on patent counts. *Environmental & Resource Economics*, 45(1): 133–155.

Lai E C and Qiu LD (2003). The North's intellectual property rights standard for the South? *Journal of International Economics*, 59(1): 183–209.

Maskus KE (2000). *Intellectual Property Rights in the Global Economy.* Washington, DC, Peterson Institute for International Economics, Peterson Institute Press.

Maskus KE (2010). Differentiated intellectual property regimes for environmental and climate technologies. Paris, OECD Publishing, May.

Maskus KE and Reichman J, eds. (2005). *International Public Goods and Transfer of Technology: Under a Globalized Intellectual Property Regime*. Cambridge, Cambridge University Press.

Mattoo A and Subramanian A (2010). Equity in climate change: an analytical review. Washington, DC, World Bank, July.

Moon S (2008). Does TRIPS Article 66(2) encourage technology transfer to least developed countries? An analysis of country submissions to the TRIPS Council (1999-2007). ICTSD Policy Brief No.2, ICTSD, Geneva.

Ocampo JA and Vos R (2008). *Uneven Economic Development*. United Nations Publications.

Patel S, Roffe P and Yusuf A, eds. (2000). *International Technology Transfer: The Origins and Aftermath of the United Nations Negotiations on a Draft Code of Conduct*. The Hague, Kluwer Law International.

Plassmann, K and Edwards-Jones, G. (2010). Carbon footprinting and carbon labelling of food products. In: Sonesson U, Berlin J and Ziegler F, eds. Environmental Assessment and Management in the Food Industry: Life Cycle Assessment and Related Approaches. Woodhead Publishing Series in Food Science, Technology and Nutrition No. 194. Cambridge, Woodhead Publishing.

Potts J, van der Meer J, Daitchman J (2010). The State of Sustainability Initiatives Review 2010: Sustainability and Transparency. Winnipeg: International Institute for Sustainable Development.

Reichman J (1996). From free riders to fair followers: Global competition under the TRIPS Agreement. *New York University Journal of International Law and Politics*, 29: 11–93.

Robins N, Clover R and Singh C (2009). A climate for recovery: The colour of stimulus goes green. London, HSBC Global Research.

Semine, N (2010). Green technologies: Opportunities for South-South trade. International Trade Forum, Issue 1. Geneva, International Trade Centre.

Seres S, Haites E and Murphy K (2009). Analysis of technology transfer in CDM projects: An update. *Energy Policy*, 37(11): 4919–4926.

Steenblik R (2006). Liberalisation of trade in renewable energy and associated technologies, biodiesel, solar thermal and geothermal energy. OECD Trade and Environment Working Papers, no. 2006/01, OECD, Paris, April.

Stern N (2007). *The Economics of Climate Change: The Stern Review*. Cambridge and New York, Cambridge University Press.

Swedish National Board of Trade (2010). The Swedish Climate Standard Project of the Swedish Ministry of Foreign Affairs. Newsletter No. 2/10. Stockholm

Tan C (2010). Confronting climate change: Towards a new international agenda for meeting the financial challenges of the climate crisis in least developed countries. Background paper for UNCTAD's *Least Developed Countries Report 2010*. Geneva, UNCTAD.

UNCTAD (2007). *The Least Developed Country Report: Knowledge, Technological Learning and Innovation for Development*. United Nations publication, sales No. E.07.II.D.8. New York and Geneva, United Nations.

UNCTAD (2009). *Trade and Development Report*, pp. 165-66. New York and Geneva, United Nations

UNCTAD (2010). *The Least Developed Countries Report 2010: Towards a New International Development Architecture for LDCs*. United Nations publication, sales No. E.10.II.D.5. New York and Geneva, United Nations.

UN/DESA. (2009a). *World Economic and Social Survey 2009: Promoting Development, Saving the Planet*. New York, United Nations.

UN/DESA (2009b). *A Global Green New Deal for Climate, Energy, and Development*. New York, United Nations.

UN/DESA, UNEP and UNCTAD (2010). *The Transition to a Green Economy: Benefits, Challenges and Risks from a Sustainable Development Perspective.* New York and Geneva. Available at: http://www.unep.org/greeneconomy/Portals/88/documents/research_products/UN-DESA,%20UNCTAD%20Transition%20GE.pdf.

UNDP (2007). *Annual Report 2007: Making Globalization Work for All*. New York

UNDP (2008). *Annual Report 2008. Capacity Development: Empowering People and Institutions.* New York.

UNEP and Bloomberg (2010). *Global Trends in Sustainable Energy Investment 2010: Analysis of Trends and Issues in the Financing of Renewable Energy and Energy Efficiency.* United Nations Environment Programme. Available at: http://bnef.com/Download/UserFiles_File_WhitePapers/sefi_unep_global_trends_2010.pdf.

UNEP, EPO and ICTSD (2010). *Patents and Clean Energy: Bridging the Gap between Evidence and Policy*. Final report. Available at www.epo.org/clean-energy

UNFCCC (1998). Kyoto Protocol. Bonn

UNFCCC (2001). Report of the global environment facility to the conference. Note by the secretariat, FCCC/CP/2001/8. Bonn.

UNFCCC (2007). Ad Hoc Working Group on Long-term Cooperative Action under the Convention, Bali Action Plan, Document FCCC/CP/2007/L.7/Rev.1. Available at : http://unfccc.int/resource/docs/2007/cop13/eng/l07r01.pdf.

UNFCCC (2008). Investment and Financial Flows to Address Climate Change: An update. Availble at http://unfccc.int/resource/docs/2008/tp/07.pdf

UNFCCC (2010). Report of the Conference of the Parties on its sixteenth session. Cancun, December. Available at: www.unfccc.int/resources/docs/2010/cop16/eng/07a01.pdf#page=2

Wooders P, Cosbey A and Stephenson J (2009). Border carbon adjustment and free allowances: Responding to competitiveness and leakage concerns. Paper presented at the Round Table on Sustainable Development, Singapore, 23 July. Paris, OECD.

World Bank (2010). Climate Technology Fund. Semi-annual operational report. Washington, DC, November.

5 NATIONAL POLICY FRAMEWORKS FOR RENEWABLE ENERGY TECHNOLOGIES

CHAPTER V

NATIONAL POLICY FRAMEWORKS FOR RENEWABLE ENERGY TECHNOLOGIES

A. INTRODUCTION

The new energy paradigm involving the greater use of RETs should be led by national governments in collaboration with the private sector, and it should be supported by a variety of stakeholders, including public research institutions, the private sector, users and consumers, on an economy-wide basis. A policy framework that strikes an appropriate balance between economic considerations of energy efficiency and the technological imperatives of RETs in developing countries and LDCs will be the cornerstone of such an agenda for change. This will necessitate two separate but related agendas. The first should ensure the integration of RETs into national policies for climate change mitigation. The second should be the steady promotion of national innovation capabilities in the area of RETs. This latter policy agenda entails addressing issues that are not only generic to the innovation policy framework, but also new issues, such as creating standards for RETs, promoting grid creation, and creating a more stable legal and political environment to encourage investments in RETs within countries.

At the national level, policymakers will need to identify market failures and opportunities in RETs, and should be able to adopt strategic policies and ensure their implementation. It has been argued that the "smart" use of policies, especially industrial policy, is important for fostering greater domestic use, production and export of RETs (see, for example, UN/DESA, 2009). Such policies will be important to steer the move away from unsustainable, carbon-intensive economic models towards more sustainable development.

Not every developing country can endeavour to attain competitive advantages in the production and innovation of RETs. However, as chapter III of this Report has stressed, simply the sustainable integration of RETs into the energy mix of countries for promoting the production or the adoption and adaptation of newer technologies requires the development of national innovation capabilities. A recent study on challenges related to RET innovation concludes that linear policies, including R&D subsidies and tax incentives, management advice and technology transfer, should be complemented by instruments that address the systemic obstacles to innovation, including the creation of institutional structures, provision of strategic information, and technology demonstration and learning platforms (Negro, Hekkert and Smits, 2008). Addressing these systemic challenges is a difficult and complex process, as pointed out in chapter III of this Report, and requires coherent policy support and coordination over the long term.

This chapter builds on the key issues presented in the previous chapters of this *TIR*. It discusses the main elements of an integrated innovation policy framework for developing countries that are seeking to use RETs while developing in an environmentally sustainable way. The term "integrated innovation policy framework" for RETs signifies addressing the issues of innovation and energy in an integrated manner. Such a policy framework would have five key functions:

(i) Defining policy strategies and goals;
(ii) Providing policy incentives for R&D, innovation and production of RETs;
(iii) Providing policy incentives for developing greater technological

The term "integrated innovation policy framework" for RETs signifies addressing the issues of innovation and energy in an integrated manner.

Such a policy framework would have five key functions.

absorptive capacity, which is needed for adaptation and use of available RETs;

(iv) Promoting domestic resource mobilization for RETs in national contexts; and

(v) Exploring newer means of improving innovation capacity in RETs, including South-South collaboration.

Defining targets is an important signal of political commitment and support...

Many of these policy incentives have been used by most of the industrialized countries, although developing countries are also increasingly using them or experimenting with their use. Keeping this in mind, the analysis seeks to contextualize the discussion to developing countries as much as possible. As the previous chapters of this *TIR* have stressed, developing countries will face different problems in RETs promotion, production and innovation, depending on their respective starting points. Nevertheless, for all developing countries, RETs present real opportunities for reducing energy poverty, and the right policies could influence the extent of benefits that could be derived from RETs use, adaptation and dissemination.

B. ENACTING POLICIES WITH RET COMPONENTS AND TARGETS

... and the policy and regulatory frameworks aimed at enforcing those targets would provide legal and economic certainty for investments in RETs.

The development of policies for RETs is complex and involves a large number of potential stakeholders. It should be approached in an integrated manner, with a long-term perspective and clearly defined roles and responsibilities. Moreover, it should include fiscal and regulatory incentives, both on the supply side (R&D and innovation) and the demand side (use and adaptation). Developing countries need to ensure that national energy policies contain targets for RE where possible, and that such targets are properly factored into innovation strategies, including mandates for on-grid and off-grid energy supply. Focusing only on either supply-side (RE provision) or demand-side management issues (such as particular user groups) may not yield as many benefits as a focus on both aspects in the long term.

RETs use and adaptation within countries requires the establishment of long-term pathways and national RE targets. These targets, although not necessarily legally binding in nature, would have to be supported by a range of policy incentives and regulatory frameworks. Defining targets is an important signal of political commitment and support, and the policy and regulatory frameworks aimed at enforcing those targets would provide legal and economic certainty for investments in RETs.

Targets may be defined in terms of requiring a specified share of REs in primary energy, and/or electricity generation. Set at regional, national and sub-national levels, they can provide a stable investment environment for project developers to operate. As noted in chapter III, a review of policy trends across countries shows a steady increase in policy activity in this regard: by 2010, more than 100 countries had introduced either a target or policy mechanism for promoting RETs. This represents a doubling of policy incentives compared with 2005. There has also been a rapid increase in the number of developing countries that are beginning to implement policies on RETs, and they now represent more than half of all the countries with such policy frameworks in place (REN21, 2010). Many new targets enacted over the past three years call for a range of 15–30 per cent of shares of renewables in energy or electricity supply by 2020. Some targets are as high as 90 per cent. Other targets require a share of renewables in primary energy or final energy supply, generally set in the range of 10–20 per cent. Developing countries now make up more than half of all countries with policy targets (45 out of 85 countries) (table 5.1).

The EU's 2007 Directive on Renewable Energy has set an ambitious target, which

Table 5.1: Renewable energy targets of selected developing economies

	Primary Energy		Final Energy	
Country/Region	**Existing share (2008)**	**Future target**	**Existing share (2008)**	**Future target**
Developing Countries				
China	9.9%	10% by 2010		15% by 2020
Egypt		14% by 2020		
Fiji				100% by 2013
Indonesia	5%	17% by 2025		
Jordan		7% by 2015 10% by 2020		
Kuwait				5% by by 2020
Lebanon				12% by 2020
Madagascar				54% by 2020
Malawi		7% by 2020		
Mali		15% by 2020		
Morocco		8% by 2012		10% by 2012
Nigeria		20% by 2012		
Pakistan		10% by 2012		
Palestine				20% by 2012
Senegal		15% by 2025		
Syria		4.3% by 2011		
Thailand		20% by 2022		
Tonga				100% by 2013
Tunisia		10% by 2011		10% by 2011
Uganda		61% by 2017		
Vietnam		3% by 2010 5% by 2020 11% by 2050		

Source: REN21 (2010).

requires that member States source an average of 20 per cent of energy from renewables. This target has resulted in intra-regional burden sharing, with some countries, such as Sweden, setting a target of 49 per cent, and others, such as Luxembourg, setting a target of as low as 11 per cent. The share of renewables in electricity supply in EU member States is expected to double to meet this target, although each member State is free to determine to what extent the power sector will contribute to meeting this share in overall energy supply.[1]

In the developing world, China adopted in 2009 a target of a 15 per cent share of renewables in final energy consumption by 2020, which is in line with EU standards, although this also includes nuclear as a non-fossil fuel source (Martinot and Li, 2010). China's RE targets are also supported by carbon intensity targets (i.e. GHG emissions relative to GDP). Other countries that have adopted RE deployment targets include Brazil, which has set a target of 75 per cent of electricity from RE sources by 2030, including through large-scale hydropower.

Targets can also be restricted to a specific RET that is more applicable in the context of a particular country. For example, India has set a target of 20 GW of solar power by 2022, while Kenya has mandated 4 GW of geothermal by 2030 (REN21, 2010). Egypt, Indonesia, Malaysia, the Republic of Korea and South Africa have also implemented some RE targets.[2] In addition, a number of sub-national governments have implemented targets for RE use, such as the state of Karnataka in India which has set a 6 GW target of renewable energy by 2015, and Jiangsu Province in China, which has set a 400 MW target for solar PV by the end of 2011.

Targets can also be restricted to a specific RET that is more applicable in the context of a particular country.

Policymakers in developing countries should also be aware of the potential trade-offs and co-benefits between low-carbon energy systems and the energy access agenda. There are some obvious advantages to be had from moving away from fossil

Dedicated incentives will also help to tip the balance in favour of RETs over conventional fossil fuel technologies in direct and indirect ways.

fuel energy generation (that is exposed to high and volatile input costs) towards the use of many existing RETs that are already competitive with conventional systems in terms of cost and reliability. For instance, small-scale RETs can often be provided at lower cost than grid extension, and they may be very important in the provision of energy to rural areas, as discussed in a previous chapter. Much progress could be made to alleviate energy poverty by focusing on rural, off-grid applications alongside efforts to establish more technologically and financially intensive grid-based RET applications. Boxes 5.1 and 5.2 provide some interesting examples of how RETs have been used in off-grid applications to promote rural energy in India.

Countries may also choose to move away from conventional energy to increase their national energy security. For example, over the past decade Chile's energy policy has focused on increasing its energy security and efficiency, as well as improving the environmental sustainability of its energy matrix (CNE, 2009). The Government accords considerable importance to diversifying its energy matrix because its entire fossil fuel supplies to meet its requirements are imported, which results in considerable price fluctuations of domestic electricity prices, depending on the international prices of oil and gas. Renewable energy sources, particularly wind power, have been included into the country's energy matrix with noteworthy results (box 5.3).

Box 5.1: Demystifying solar energy in poor communities: The Barefoot College in action

The Barefoot College is a non-governmental organization (NGO) that was established in 1972 in the state of Rajasthan, India. Its goal is to empower rural communities by making them self-sufficient in sustainable energy supply through the application of different solutions and approaches, one of them being the adoption of solar technology to electrify rural and remote villages. The College has pioneered solar electrification in rural areas since 1984. Since then it has contributed to electrifying villages across 16 states of India and 17 countries in Africa, Asia and South America, which represents a total solar energy generating capacity of more than 800 kWp (kilowatt-peak).

The mission of the College is "the demystification of solar technology and decentralisation of its application". This means giving responsibility for the fabrication, installation, utilization, repair and maintenance of sophisticated solar lighting units to rural and often illiterate and semi-literate men and women. It believes that people from rural or poor backgrounds do not need formal educational qualifications to acquire skills that can be of service to their community. Therefore the training of local inhabitants, so- called barefoot solar engineers (BSEs), is an essential part the Barefoot College programme. The BSEs are responsible for installing, repairing and maintaining the necessary equipment for a period of at least five years, as well as for setting up a rural electronic workshop where components and equipment needed for the repair and maintenance of solar units are stored. The programme has received funding from the European Commission, the Asian Development Bank (ADB), United Nations Development Programme (UNDP) and the Indian Ministry of Non-conventional Energy Sources.

Under the Barefoot College approach, every family that wishes to obtain solar lighting must pay an affordable contribution every month, irrespective of how poor it is. This is considered important for creating a sense of ownership and responsibility. The monthly fee to be paid by each electrified household is determined by how much each family spends on kerosene, candles, torch batteries and wood for lighting every month. A percentage of the total contributions made by the households helps to pay the monthly stipend of every BSE, and the remainder covers the costs of components and spare parts.

The process of solar electrification is not undertaken until the villagers have expressed a desire for solar lighting and agree, in writing, to pay or collect the nominal monthly fee, select BSEs for training and arrange for a rural electronic workshop to be set up. Barefoot College implements this agreement on behalf of the rural community, initiating and ensuring complete participation.

Several organizations, including the ADB, UNDP, Skoll Foundation (United States), Fondation Ensemble (Together Foundation) (France), Het Groene Woud (The Green Forest) (Holland) and the Indian Technical and Economic Cooperation Division of the Ministry of Foreign Affairs have supported its replication in Afghanistan, Bolivia, Bhutan, Cameroon, Ethiopia, Gambia, Malawi, Mali, Mauritania, Rwanda, Sierra Leone and the United Republic of Tanzania.

Sources: UNCTAD, based on Barefoot College website at: http://www.barefootcollege.org/.

Box 5.2: Lighting a billion Lives: A success story of rural electrification in India

The Energy and Resources Institute (TERI), an international think tank based in India, launched an ambitious global initiative called Lighting a billion Lives (LaBL) in February 2008. It seeks to provide energy-impoverished communities around the world with access to clean sources of lighting through solar technologies, and to replace kerosene/paraffin lamps with clean and environment-friendly solar lanterns.

The LaBL initiative is based on an entrepreneurial model of energy service delivery aimed at providing high-quality and cost-effective solar lanterns that are disseminated through solar charging stations set up in non-electrified or poorly electrified villages. The charging station is operated and managed by a local entrepreneur trained under the initiative, who charges the rural inhabitants a small, affordable fee for renting the solar lanterns every evening. This fee-for-service model ensures that the poorest socio-economic groups have access to clean energy at an affordable price. While the capital cost of setting up the charging station in the village is raised by TERI through government agencies, corporate donors and communities, the operation and maintenance costs are borne by the users of the solar lanterns in the form of the rent that they pay to the operator of the charging station.

Since its launch in 2008, the initiative has been replicated successfully across 640 villages in 16 states of India, improving the lives of more than 175,000 people. So far, around 35,000 rural households in India have replaced kerosene lamps with clean and environment-friendly solar lanterns. The project has provided "green" livelihood opportunities to 700 rural people who have become operators of the solar charging stations, with monthly earnings of 3,000–4,000 Indian rupees from renting out the solar lanterns in their villages. Furthermore, the initiative is currently saving more than 100,000 litres of kerosene every month (based on an estimate of 3 litres per household/per month) that was previously being used by the beneficiaries for lighting. This translates into a mitigation of around 400 tons of CO2 every month.

The initiative has also formed a basis for South-South cooperation through structured training and capacity-building programmes, technology transfer initiatives, piloting of successful delivery models and identification of local partners for replicating and scaling up the model in various developing countries. In collaboration with the United States Agency for International Development (USAID), high-quality solar lanterns developed under the initiative are being sent to Afghanistan. In addition, the fee-for-service delivery model is being piloted in Cameroon, the Central Africa Republic, Kenya and Malawi, in collaboration with UN-Habitat, and in Bangladesh and Sierra Leone with UNIDO support. Furthermore, South-South cooperation is being established under the ADB's Energy for All partnership through capacity-building and training-of-trainer programmes in Cambodia, Indonesia and the Philippines. The initiative is also being expanded in Uganda through local partnerships developed in collaboration with private distribution networks in the country.

At the national policy level in India, the LaBL initiative has caught the attention of the central Government. Its delivery model has been adopted by the Government of India's Jawaharlal Nehru National Solar Mission initiative, and is being promoted to enhance access to clean energy in remote energy-impoverished regions of the country. In addition, the solar charging stations are also being used to provide a mobile telephone recharge facility under the Government's programme to provide last-mile connectivity to rural and remote communities, thus creating linkages between different development-oriented initiatives.

Source: TERI for *Technology and Innovation Report 2011*.

C. SPECIFIC POLICY INCENTIVES FOR PRODUCTION AND INNOVATION OF RETs

The successful development and deployment of any technology, especially relatively new ones such as RETs, needs the support of several dedicated institutions responsible for the different technical, economic and commercialization aspects. Such support can be organizational (dedicated RET organizations) or it can take the form of incentives to induce the kinds of behaviour required to meet the targets set for RETs. The dedicated incentives will also help to tip the balance in favour of RETs over conventional fossil fuel technologies in direct and indirect ways, as described in chapter III. The most direct impact of such dedicated incentives for RETs will be their fostering of technological activities for developing state-of-the-art RETs, and promoting their adoption and utilization in countries. An indirect effect of such incentives would be to encourage innovators to engage in incremental technologi-

Box 5.3: Increasing energy security through wind power in Chile

Chile imports its entire requirement for fossil fuels, which causes domestic electricity prices to fluctuate rapidly, depending on the international prices of oil and gas. During the past decade, Chile's energy policy has therefore focused on increasing its energy security and efficiency, as well as improving the environmental sustainability of its energy matrix (CNE, 2009). Diversification of its energy matrix through the inclusion of RE sources, particularly wind power, is also considered very important. Indeed, this has produced noteworthy results. Installed wind generation capacity in Chile increased rapidly in 2009, rising from less than 20 MW in 2007 to exceed 162 MW in 2010 (box table 5.3.1).

Box table 5.3.1 Installed wind generation capacity in Chile, 2011 (MW)

Owner	Power station's name	Year of initial operation	No. of units	Total net power (MW)
Edelaysen	Central Eolica Alto Baguales	2001		1.98
ENDESA	Canela	2007	11	18.0
ENDESA	Canela 2	2009	40	59.4
Eolica Monte Redondo	Monte Redondo	2010	19	37.6
Norvind	Totoral	2010	23	45.5
Total				162.5

Source: National Energy Commission, Chile, at: www.cne.cl/cnewww/opencms/.

Three factors may explain the strong growth of the wind energy market in Chile. First, several areas along Chile's long coastline have been identified as having the potential to generate wind powered electricity due to abundant availability of wind (Jara, 2007). Second, an appropriate legal and institutional framework has been put in place to facilitate the growth of RE. For instance, a law (Ley Corta) was enacted in 2004 to remove some of the barriers that hindered the introduction of RE in the country. This was followed by a new law (Ley No. 20.257) in 2008 to force electricity producers with more than 200 MW of installed capacity to obtain or generate at least 5 per cent of their sales from renewable sources. This requirement will increase by 0.5 per cent annually as of 2015 until it reaches 10 per cent in 2024. And third, the prospects of high future energy prices have provided additional investment incentives for wind power generation in Chile (LAWEA, 2010).

Overall, Chile's prospects in the wind energy market are promising. In fact the country is considered, along with Brazil and Mexico, as one of the Latin American leaders in the wind energy market. Recent estimates forecast that Chile will hold this position at least until 2025 (Emerging Energy Research, 2011).

Source: UNCTAD, based on CNE (2009), Jara (2007) and Emerging Energy Research (2011).

cal improvements in RETs that lead to cost reductions in their manufacture, thereby making them more affordable for a larger segment of the population in developing countries. Technological improvements could also be aimed at improving processes for the disposal of RETs at the end of their lifespan and for newer uses of RETs in economic processes. For example, there is a huge demand for the dual use of solar batteries, for lighting and for charging mobile phones in rural areas.

Such innovation capabilities could be encouraged by a number of mechanisms and by the provision of incentives within the integrated innovation policy framework for RETs. Some incentives are somewhat similar to those offered in several other sectors, such as public research grants, funding schemes, the establishment of technology clusters and special economic zones (SEZs), to promote production and collaboration, and to encourage public-private partnerships (PPPs). Incentives for the production and innovation of RETs could also be formulated as part of national energy policies, as discussed in more detail below.

1. Incentives for innovation of RETs

Incentives for innovation, production and R&D are granted to promote risk-taking by the private sector, to improve domestic capacity to engage in learning activities, and to promote basic and secondary research in the

Box 5.4: Public support for RETs in the United States of America and China

In the United States, public support initiatives for early-stage technologies are funded through the Department of Energy's Advanced Research Project Agency-Energy (ARPA-E). The current stimulus package provided over $400 million in 2009-2010, with awards ranging from between $0.5 and $10 million per annum. Spanning over three funding cycles, the RE projects funded include those associated with biomass, direct solar fuels, electro-fuels, vehicle technologies and grid storage applications. Grant recipients have been primarily universities, R&D departments in large private corporations and companies pursuing individual early-stage technologies. Many of the companies have gone on to raise venture capital finance, which demonstrates that grant provision can leverage private sector investment. The American Energy Innovation Council has called for a massive scale-up of funding for the APRA-E programme from existing levels to $1 billion per annum (Cleantech Open, 2010).

Among developing countries, China has several projects in carbon-capture technology as a way to offset rising emissions from indigenous coal-based power production. The Luzhou Natural Gas Chemicals plant is already capturing CO2 on a commercial scale for use in urea production. China and Japan are jointly cooperating on the Harbin Thermal Power Plant in Heilungkiang Province. The plant will capture CO2 and transport it to the Daqing Oilfield approximately 100 km away where it will be injected into a disused reservoir. Other projects under development include the GreenGen IGCC Coal Power Plant in Tianjin, and the Shanghai Shidongkou ultra-critical power plant.

Source: UNCTAD.

public sector. Some of the policy incentives listed here are aimed specifically at the private sector, such as green economic clusters and SEZs, to boost enterprise activity, whereas others are hybrid instruments that can be granted to promote both public and private sector activity, such as collaborative PPPs. Others, such as public research grants, are offered primarily to the public sector.

a. Public research grants

The most common form of government support for supply-side RET innovation is the provision of government grants to universities and R&D centres for early stage development. Increasingly, grants are being offered for the development of low-carbon energy technologies, rather than for all RETs. They take into account the potential for decarbonizing existing fossil fuel technologies through carbon capture and storage (CCS), or they seek to promote the use of nuclear energy as a low-carbon alternative. Both of these have the potential to reduce growth in emissions from coal plants in rapidly industrializing countries. In early 2009, both the EU and the United States announced significant funding for R&D in CCS technologies and for demonstration projects (box 5.4). Committed funding in the United States for early-stage deployment is currently $4.3 billion (box 5.4), while carbon credits set aside specifically for CCS in the EU could total over €12 billion by 2014 (ZEP, 2008).

b. Grants and incentives for innovation of RETs

Most industrialized countries have used grant-support schemes to promote the use of low-carbon or renewable energy resources in the development and early-stage deployment of RETs for electricity, heat and transport, and for installing more energy-efficient power-generation plants. These grant schemes are usually competitive in nature, with governments seeking to maximize returns (box 5.5).

In some developing countries, innovative mechanisms are being developed to reduce the risk to technology developers in designing and delivering more cost-effective and resilient RETs.

In some developing countries, innovative mechanisms are being developed to reduce the risk to technology developers in designing and delivering more cost-effective and resilient RETs. Examples and experiences of countries (e.g. as described in box 5.2) help to demonstrate an important policy finding: not all countries need to provide large grants. In countries that have significant resource constraints, incentive structures can target small-scale projects designed to encourage private sector involvement in RET development and deployment in small rural settings. An example is the Africa Enterprise Challenge Fund's Renewable Energy and Adaptation to Climate Technologies (AECF REACT) programme (box 5.6).

Box 5.5: Examples of grant schemes in industrialized countries

In the EU, the Seventh Framework Research Programme (FP7) for the period 2007–2013 seeks to strengthen the European industrial technology base while encouraging international competitiveness in RETs. The energy component within this major regional programme is worth €2.35 billion. Competitive grants provided to commercial companies cover up to 50 per cent of their costs, and those to non-commercial bodies cover up to 75 per cent of costs. Demonstration projects receive up to 50 per cent of costs. Research is funded across a range of technologies, including hydrogen and fuel cells, the use of renewables for electricity generation, heating and cooling, and renewable fuel production.[a]

In Australia, the Renewable Energy Demonstration Program (REDP) is the centerpiece of the Australian Centre for Renewable Energy (ACRE), which provides substantial grants to fund large-scale, grid-connected RET demonstration projects across a range of sectors and technologies. The grants cover expenditure of up to a third of total costs. The REDP has funded four large projects, including two geothermal projects (amounting to 152 million Australian dollars), one ocean energy project (66 million Australian dollars) and one integrated project (15 million Australian dollars). The Australian Solar Institute, which provides similar grant support for solar thermal and PV technologies, had disbursed more than 44 million Australian dollars to 13 projects by mid-2010.[b]

The Canadian Clean Energy Fund also seeks to support RETs demonstration projects, both on- and off-grid, including smart grid concepts, electrical and thermal energy storage, hybrid systems (including those with limited fossil fuel input), marine energy, solar PV, solar thermal, very low head hydro and in-stream river current systems, geothermal and bio-energy. Established in 2010, the fund will provide 850 million Canadian dollars, of which 80 per cent will be for CCS and 200 million Canadian dollars for smaller scale RE projects. There is an additional 150 million Canadian dollar R&D grant facility. The fund aims to support systems demonstration that includes codes, permits and grid connections, as well as those that reduce capital costs and demonstrate reliability. Funds are provided for up to 50 per cent of the total demonstration costs up to a limit of 50 million Canadian dollars per project.[c]

Source: UNCTAD.

[a] cordis.europa.eu/fp7/home_en.html.

[b] www.ret.gov.au/energy/energy%20programs/cei/acre/redp/Pages/default.aspx.

[c] www.nrcan.gc.ca/eneene/science/ceffep-eng.php.

Box 5.6: The Renewable Energy and Adaptation to Climate Technologies Programme of the Africa Enterprise Challenge Fund (AECF REACT)

This programme seeks to catalyse private sector investment and innovation in low-cost, clean energy and climate change technologies, particularly those aimed at rural populations that have limited or no grid access. It operates on the basis of a "challenge fund", managed by a private sector fund manager, which can be accessed through competitive applications by for-profit RET companies. Successful applicants receive grants and interest free repayable grants up to a maximum of $1.5 million. Applicant companies are required to match the AECF's REACT funding with an amount equal to or greater than 50 per cent of the total cost of the project. As AECF's REACT aims to leverage its funds, those applications where companies contribute a greater percentage of matched funds, or that incorporate a greater percentage of the repayable grant, are given more favourable consideration.

The World Bank Group has taken similar approaches to reduce the risk to private sector developers in adapting existing technologies to developing-country environments. One such project is the International Finance Corporation's Fuel Cell Financing Initiative for Distributed Generation Applications (IFC FCFI). This IFC/Global Environment Facility-funded project seeks to promote the utilization of fuel cells in stationary power applications. It targets a range of potential technologies that could provide efficiencies of up to 90 per cent. As part of this project, contracts have been signed with IST holdings (South Africa) and Plug Power Inc. (United States) in 2005, with the intention of installing 400 fuel cells in remote locations across South Africa.

Source: UNCTAD, based on AECF (www.aecfafrica.org/react/index.php) and IFC.
(http://www.thegef.org/gef/sites/thegef.org/files/repository/Global-FuelCellsFinancingInit-DistribGeneration.pdf).

Box 5.7: Lighting Africa

Lighting Africa is a joint initiative of the World Bank and the IFC. Its goal is to provide safe, affordable and modern off-grid lighting using RETs to 2.5 million people by 2012, including a target of 250 million people by 2030 in sub-Saharan Africa. The initiative targets rural, urban and peri-urban populations who lack access to electricity, especially low-income households and businesses. As mentioned earlier in this Report, it is estimated that energy poverty is especially severe in sub-Saharan Africa, with the region accounting for 500,000 of around 1.7 billion people worldwide who live without electricity. Rural electricity access rates in the region are as low as 2 per cent, which hinders social and economic development.

In order to achieve its goals, the programme works with product manufacturers and distributors, consumers, financial institutions, development partners and governments to help build markets for reliable off-grid lighting products. Lighting Africa's strategy is based on four pillars: (i) facilitating consumer access to a range of affordable and reliable products and services; (ii) catalysing the private sector by strengthening the ties between the different actors to provide lower cost products; (iii) improving market conditions, by removing technical, financial, policy and/or institutional barriers; and (iv) mobilizing the international community to promote delivery of modern lighting services to the poor in Africa.

Despite its relatively recent creation in September 2007, the initiative has already achieved significant results: over 190,000 portable solar lamps, which had passed Lighting Africa quality tests, have been sold in Africa, providing more than 950,000 people with cleaner, safer, better lighting and improved energy access. So far, eight products have passed Lighting Africa quality tests, and are available in the African market at prices ranging between $22 and $97. Since February 2011, the first testing laboratory in East Africa has been offering testing of off-grid lighting products as a commercial service to manufacturers and distributors. The laboratory, located at the University of Nairobi, uses Lighting Africa's low-cost initial screening method. The governments of Ethiopia, Mali and Senegal have signed agreements with Lighting Africa to integrate lighting services into their rural energy programmes.

In addition, Lighting Africa has established the Lighting Africa Outstanding Products Awards which provide increased recognition and visibility to particularly good off-grid lighting products in different categories: best room lighting performance, best task lighting performance, best portable torch light, best economic value and top performer overall.

Source: UNCTAD, based on Lighting Africa, at: www.lightingafrica.org/.

Other examples of similar smaller scale projects that have achieved significant results abound (see, for example, box 5.7).

c. Collaborative technology development and public-private partnerships

Several OECD governments have established PPPs to promote the commercialization and deployment of RETs. Providing public funding for long-term technology collaboration, together with private sector technology know-how, may result in more effective innovation in RETs. The European Commission (EC) has recognized that markets and energy companies acting alone are unlikely to deliver the technological breakthroughs quickly enough to meet climate change and RET policy goals. Owing to locked in investments, vested interests in existing technology models and potentially large investment risks, progress is likely to be slow without some form of PPP (EC, 2010). To accelerate the commercialization and deployment of RETs, the EC has developed a series of RET roadmaps for key sectors that promote PPPs in this area (box 5.8).

Some developing countries, such as Bangladesh, are also actively assessing the benefits of PPP structures for both R&D and deployment of RETs. They have been exploring PPP solutions across a range of other sectors, including energy and health. Promoting such models in the energy sector may complement national efforts to encourage growth of innovation capacity and energy security.

Providing public funding for long-term technology collaboration, together with private sector technology know-how, may result in more effective innovation in RETs.

d. Green technology clusters and special economic zones for low-carbon technologies

National and sub-national governments in both developed and developing countries are increasingly providing targeted support to encourage critical mass in low-carbon manufacturing and RET cluster centres. Recognizing the rapid growth in demand for RETs, governments are competing

Box 5.8: Examples of public-private partnerships

The United Kingdom has created the Energy Technologies Institute (ETI) – a PPP between the Government and a number of multinational companies: BP, Caterpillar, EDF, E.ON, Rolls-Royce and Shell. Each of these companies has a seat on the board. The Government is committed to providing funds of £50 million annually over a period of 10 years starting in 2008-2009. Matching funds are to be raised by the Energy Research Partnership, a government entity created for such a purpose. The ETI is tasked with developing technologies that will help the United Kingdom meet its legally binding 2050 carbon reduction targets under the Climate Change Act. It funds projects that deliver sustainable and affordable energy for heat, power transport and associated infrastructure using a range of technology options. The Institute seeks to demonstrate technologies and develop the skills base and necessary supply chains for the required level of technology deployment during the period 2020–2050. It is not a grant-awarding body; rather, it makes targeted investments in large-scale engineering projects that may have a strategic impact on the country's economy.

In the United States, the Department of Energy's Energy Frontier Research Center Program is also adopting a PPP approach by bringing together corporations, national laboratories and universities. In 2009, 46 research centres were established, each with an annual funding of $2–$5 million as part of this programme. Energy innovation hubs have also been established to address specific technological challenges. These include the Energy Efficient Building Design Hub, led by Pennsylvania State University, and the Fuels from Sunlight Energy Innovation Hub led by the Joint Center for Artificial Photosynthesis. There are concerns that the ability of such PPPs to attract long-term private sector finance may suffer unless there is more ambitious legislation on climate mitigation and REs.

Source: UNCTAD, based on Cleantech Open (2010).

to secure investment in R&D and manufacturing, both from developed-country manufacturers seeking to scale up production in lower cost markets, and from emerging-economy manufacturers in India and China.

For developing countries, low-carbon SEZs and green clusters may be useful measures.

For developing countries, low-carbon SEZs and green clusters may be useful measures for enhancing industrial competitiveness and FDI, especially for boosting the private sector. They can be used as a means to diversify economic activity while maintaining protective barriers, and to pilot new policies and approaches. Where there are successes, these can feed into wider innovation policy and set benchmarks for the development of domestic industry. Environmental and efficiency standards developed within an SEZ can be taken up by governments and applied at national and/or regional levels.

These clusters typically provide suitable infrastructure, skills and proximity to markets. Many countries where RET clusters have been successfully deployed have already set ambitious domestic carbon reduction goals and created supportive regulatory environments.

In China, the idea of a low-carbon SEZ was first proposed in 2007, with support from the Government of the United Kingdom and China's National Development and Reform Commission. An initial pilot low-carbon SEZ is being set up in the industrial province of Jilin. In India, the Ministry for New and Renewable Energy has announced support for manufacturing of RETs through the creation of a dedicated SEZ in the city of Nagpur. It will focus on strategically important input materials, process and testing equipment, devices and systems components. The ministry is offering to facilitate joint ventures and technology transfers to achieve this as part of an overall package of incentives for investment in this sector. Other measures include rationalizing the customs and excise duty structure, liberalizing import regulations, and providing income tax concessions and concessional financing.

Masdar, a venture in Abu Dhabi is another example which is positioning itself as an R&D hub for new energy technologies to drive the commercialization and adoption of these and other technologies in sustainable energy, carbon management and water conservation. In the United States, cities such as Seattle and Boston have been suggested as potential clean-tech innovation hubs. In addition, a number of universities are supporting renewable technology business incubators, such as the New York City Accelerator for a Clean and Renewable Economy.

2. Innovation and production incentives and regulatory instruments in energy policies

As noted earlier in this chapter, a fundamental characteristic of integrated innovation policy frameworks for RETs is that they promote interactions between general innovation policy incentives and energy policies of countries. Ongoing reforms in the energy sector of most developing countries offer a good opportunity to establish regulatory instruments and production obligations geared towards promoting investment in RETs and energy production based on these technologies. Many regulatory instruments are available, including assessment and auditing, benchmarking, mandates, monitoring, standards and quota systems, and associated tradable certificates/permits (IEA, 2011; Komor and Bazilian, 2005; Oliver et al., 2001; Schaeffer and Voogt, 2000). Production incentives can take the form of financial incentives for installed RE capacity to be used, reducing the risk of investing in and using RETs to provide energy services by increasing the rate of return, and reducing the payback period (van Alphen, Kunz and Hekkert, 2008).

a. Quota obligations/renewable portfolio standards

Quota obligations or renewable portfolio standards (RPSs) are hybrid economic/regulatory mechanisms that mandate electricity providers to supply a specific minimum amount of electricity generated from RE sources by a set target date. These instruments have been used in many countries to accelerate the transition to RE systems and to achieve the same outcomes as feed-in tariffs. The additional costs of meeting the quota are usually passed through to consumers. By 2010, RPSs had been introduced by at least 10 national governments and 46 sub-national bodies globally (table 5.2). Most obligations required a RE-generated electricity component of

Ongoing reforms in the energy sector of most developing countries offer a good opportunity to establish regulatory instruments and production obligations...

...geared towards promoting investment in RETs and energy production based on these technologies.

Table 5.2: Countries/states/provinces with RPS policies

Year	Cumulative No.	Countries/states/provinces added that year
1983	1	Iowa (United States)
1994	2	Minnesota (United States)
1996	3	Arizona (United States)
1997	6	Maine, Massachusetts, Nevada (United States)
1998	9	Connecticut, Pennsylvania, Wisconsin (United States)
1999	12	New Jersey, Texas (United States); Italy
2000	13	New Mexico (United States)
2001	15	Flanders (Belgium); Australia
2002	18	California (United States); Wallonia (Belgium); United Kingdom
2003	21	Japan; Sweden; Maharashtra (India)
2004	34	Colorado, Hawaii, Maryland, New York, Rhode Island (United States); Nova Scotia, Ontario, Prince Edward Island (Canada), Andhra Pradesh, Karnataka, Madhya Pradesh, Orissa (India); Poland
2005	38	District of Columbia, Delaware, Montana (United States); Gujarat (India)
2006	39	Washington State (United States)
2007	44	Illinois, New Hampshire, North Carolina, Oregon (United States); China
2008	49	Michigan, Ohio (United States); Chile; the Philippines; Romania
2009	50	Kansas (United States)

Sources: Reproduced from IEA, Global Renewable Energy Policies and Measures database; REN21 (2010).

Note: Cumulative number refers to number of jurisdictions that had enacted RPSs in a given year. Jurisdictions are listed under year of first policy enactment; many policies were revised in subsequent years. Six Indian states (Haryana, Kerala, Rajasthan, Tamil Nadu, Uttar Pradesh, and West Bengal) are not shown in the table since the year is uncertain.

Box 5.9: Renewable portfolio standards in the Philippines

In the Philippines, RE is steadily becoming a greater part of the energy portfolio. In terms of installed capacity, the Philippines is currently second in the world for geothermal and third for biomass power (REN21, 2009). In 2009, RE sources accounted for 34 per cent of total installed capacity (Almendras, 2010). However, a number of problems relating to RE have emerged. For instance, some commercial wind turbines have been disabled and their components made of valuable materials, such as copper and aluminum, are sold on the black market. Also, incumbent transmission and distribution companies have been able to charge higher transmission rates for wheeling power from renewable resources (Sovacool, 2010). The Government has taken steps to address these issues and continue the promotion of RE through the Renewable Energy Act in 2008.

This Act includes mandates for on-grid and off-grid, and addresses general issues relating to the provision of energy through an RPS. For on-grid and off-grid suppliers, the newly created National Renewable Energy Board will set minimum required quotas for sourcing from REs, thereby contributing to RE growth. A number of mechanisms have been created as incentives for stakeholders to invest in RE, including feed-in tariffs that give priority to RE systems for connections to grid and the purchase of this electricity by grid operators, as well as a fixed tariff for each type of RE for no less than 12 years. An RE certification process has also been created. In general, RE suppliers are entitled to an income tax holiday for the first 7 years of operation and duty-free importation of equipment for the first 10 years, although these are subject to a number of conditions. Also, consumers have the option of purchasing renewable power from suppliers (Government of the Philippines, 2008).

Sources: REN21 (2009); Government of the Philippines (2008); Sovacool (2010) and Almendras (2010).

Renewable portfolio standards can act as a powerful tool for RETs promotion.

between 5 per cent and 20 per cent, with targets usually extending to 2020 and beyond (REN21, 2010). As the table shows, developing countries such as Chile, China, India and the Philippines have also introduced RPSs.

Renewable portfolio standards can act as a powerful tool for RETs promotion, since they can be accompanied by regulations that force electricity distributors to disclose the mix of fuels and related emissions in their power supply. In the United Kingdom, for example, the Renewables Obligation introduced in April 2002 is the main policy mechanism which aims at increasing RE deployment. The policy obliges electricity suppliers to source an increasing share of their electricity from REs. It contains a penalty structure that can be invoked when the renewables obligation is not met. The obligation to source renewables is a moving target within the policy: as of 2002-2003, it required that a minimum of 3 per cent of electricity supplied be sourced from renewables, and this share is set to increase to 15.4 per cent in 2015. A tradable certificate called a Renewables Obligation Certificate (ROC) is issued for each MWh of electricity produced. Electricity suppliers can meet their obligation either through their own power generation, purchase certificates from other generators, pay a buy-out penalty, or a combination of the above. The ROC system has been under constant development since its introduction in 2002. As of 2009, a conscious policy decision was made to encourage technologies that were less developed by providing higher levels of financial support that such technologies required. It also sought to ensure that more mature technologies, such as onshore wind, were not being overcompensated. In practice, different technologies received a different number of ROCs per MWh produced. Key to the ROC's success has been the setting of long-term time frames. The scheme was recently extended to 2037. While the mechanism itself may have been relatively efficient, the actual deployment of renewable generation capacity has been hampered somewhat by planning delays and by issues related to grid connection. The United Kingdom's energy regulator, Ofgem, has estimated that the Renewables Obligation cost the average household in the country £7.35 per annum in 2007 (approximately £200 million), and has forecast that this will rise to £11.41 by 2010–2011 (Scottish Executive, 2009).

In May 2003, Sweden introduced a system of electricity certificates in order to meet its targets for the production of electricity from RE sources. Since its introduction in 2003, the policy objective of the legislation has been expanded to include the production of electricity from peat as a fuel in combined heat and power plants. With effect from 1 January 2007, the policy target includes an increase in the production of electricity from renewable sources by 17 TWh by 2016, relative to 2002. This system has also been extended to 2030.

Some developing countries, such as the Philippines, are also using RPS (box 5.9).

b. Feed-in tariffs

The most common form of a guaranteed fixed price system is the feed-in tariff (FIT), which offers a price incentive to investors. Governments determine the price per kWh that the local distribution company will have to pay for power generation from renewables that is fed into the local distribution grid. The costs can be financed through a levy on electricity applied to all consumers. Tariffs vary widely between countries, and even within countries, according to the technology used, time (e.g. peak or base load tariffs) and seasons. FITs are agreed as part of a power purchase agreement (PPA). Standard, reliable, long-term PPAs offer a clear guarantee to the private sector and their financiers that they can hook up their power plant to the grid and receive a certain payment for energy over a set period of time (Oliver et al., 2001).

The most common form of a guaranteed fixed price system is the feed-in tariff (FIT), which offers a price incentive to investors.

Table 5.3: Countries/states/provinces with feed-in tariff policies[a]

Year	Cumulative No.[b]	Countries/states/provinces added that year
1978	1	United States
1990	2	Germany
1991	3	Switzerland
1992	4	Italy
1993	6	Denmark; India
1994	8	Spain; Greece
1997	9	Sri Lanka
1998	10	Sweden
1999	13	Portugal; Norway; Slovenia
2000	13	—
2001	15	France; Latvia
2002	21	Algeria; Austria; Brazil; the Czech Republic; Indonesia; Lithuania
2003	27	Cyprus; Estonia; Hungary; the Republic of Korea; Slovakia; Maharashtra (India)
2004	33	Israel; Nicaragua; Prince Edward Island (Canada); Andhra Pradesh and Madhya Pradesh (India)
2005	40	Karnataka, Uttarakhand and Uttar Pradesh (India); China; Turkey; Ecuador; Ireland
2006	45	Ontario (Canada); Kerala (India); Argentina; Pakistan; Thailand
2007	54	South Australia (Australia); Albania; Bulgaria; Croatia; the Dominican Republic; Finland; Mongolia; The former Yugoslav Republic of Macedonia; Uganda
2008	67	Queensland (Australia); California (United States); Chattisgarh, Gujarat, Haryana, Punjab, Rajasthan, Tamil Nadu and West Bengal (India); Kenya; the Philippines; the United Republic of Tanzania; Ukraine
2009	77	Australian Capital Territory, New South Wales, Victoria (Australia); Japan; Serbia; South Africa; Taiwan Province of China; Hawaii, Oregon and Vermont (United States).
Early 2010	78	United Kingdom

Sources: Reproduced from REN21 (2010).

Note: [a] *Many policies have been revised or reformulated in years subsequent to the initial year shown for a given country. For example, India's national feed-in tariff from 1993 was largely discontinued, but new national feed-in tariffs were enacted in 2008.*

[b] *Cumulative number refers to the number of jurisdictions that had enacted a feed-in policy by the given year, but policies in some countries were subsequently discontinued. The number of existing policies cited in REN21 (2010) is 75.*

Box 5.10: Feed-in tariffs for biogas and solar PV, Kenya

The Government of Kenya has introduced a special FIT for electricity. Geothermal power generators will receive 8.5 cents (around 6.60 Kenyan shillings) per kWh, and wind power producers and biomass producers will receive 12 cents and 8 cents respectively. The FIT was launched in 2008 in order to provide an incentive for RE-sourced power generation and was revised in 2010 to include geothermal power. The rates for wind and biomass were also raised. Power producers sell electricity to the Kenya Power and Lighting Company at a predetermined fixed tariff for a certain period of time. The feed-in tariff was again revised in January 2010 to include biogas and solar PV sources of electricity generation. Kenya is the regional leader in the solar market, with an installed capacity of 4 MW. The technology benefits 200,000 rural homes and 25,000–30,000 photovoltaic modules have been sold so far.

Source: UNCTAD.

There is limited experience of FIT implementation in developing countries, but there has been a recent upsurge over the past five years.

Feed-in tariffs generally have been very effective, and experience in developed countries shows that this policy instrument has resulted in a substantial increase in capacity of RE-based power systems (van Alphen, Kunz and Hekkert, 2008; IEA, 2011; Komor and Bazilian 2005; Schaeffer and Voogt, 2000).[3] Table 5.3 shows the number of countries/states/provinces around the world that have adopted FIT policies to date.

There is limited experience of FIT implementation in developing countries, but there has been a recent upsurge over the past five years. For example, in Algeria, the price of energy generated by hydro, waste, wind and solar PV/concentrated solar power includes a renewable premium calculated as a percentage of the average price of electricity. In Ghana, Botswana, Swaziland, South Africa and the United Republic of Tanzania, plans for FITs are being developed for a range of technologies. Mauritius has had an FIT for bagasse cogeneration since 1957, and there are plans to extend this to wind, solar and hydro. Many other countries are in the process of developing FITs (Curren, 2010), and there have been special efforts to make FITs appropriate to developing-country contexts (Moner-Girona, 2009; and see box 5.10 for FITs in Kenya).

Flexibilities in the TRIPS Agreement can be used to mitigate adverse effects of IPRs.

3. Flexibilities in the intellectual property rights regime

Much has been written about the various flexibilities contained in the WTO Agreement on Trade-Related Aspects of Intellectual Property Rights (TRIPS Agreement) that can be used to mitigate adverse effects of IPRs on reverse engineering and on incremental innovation in developing countries, both of which are important for technological learning. These flexibilities are discussed below.

(i) *Patentability criteria*. Since the three prerequisites of novelty, industrial applicability/utility and inventive step are not defined under Article 27 of the TRIPS Agreement, hence national patent regimes set different standards that need to be met by inventors.[4] A lax standard (low level) for 'inventive step' can result in a proliferation of patents over a given technology, whereas a stringent standard implies that improvements that are not significant cannot be accorded a patent right. It has been argued that setting a high level for inventive step allows firms in developing countries to engage in incremental innovations, since these will not be allowed to be patented within their domestic contexts.

(ii) *Exceptions to granted patent rights*. In some sectors, such as public health, two important exceptions can be made. One is the experimental use exception, which allows universities and public sector institutions to use patented products for research purposes, and in some countries this has also been extended to firms. The second is the regulatory review exception. The possibility of its application to RETs remains to be explored.

(iii) *Parallel imports*. According to Article 6 of the TRIPS Agreement, WTO members are free to authorize or to exclude parallel imports of IPR-protected goods.

(iv) *Compulsory licences*. The TRIPS Agreement authorizes the granting of compulsory licences without limiting the substantive grounds for such grant. As discussed in chapter IV, it has been proposed in several international forums that compulsory licensing could be used to promote the goal of greater access to and diffusion of RETs in developing countries.[5]

(v) *Competition law and policy*. By effectively controlling an abuse of dominant positions, such as the unjustified refusal by the patent holder to license an invention for the purpose of extending monopoly power to a secondary market not covered by the IPR, competition law and policy may make important contributions to the design of an IP system that appropriately balances incentives for originators and the promotion of follow-on innovation.

4. Applicability of policy incentives to developing countries

The policy incentives discussed above are highly relevant for promoting innovation capabilities in developing countries. Several of these mechanisms, such as public research grants, green clusters, SEZs and collaborative partnerships, have worked well in various countries when applied to other technologies. Depending on the level of development of a country, parameters for implementation and support may need to be nuanced.

The two incentives related to the energy sector presented above can be useful for encouraging new, cost-effective innovations in RETs by both the private and public sector. Of the two, quotas/renewable portfolio standards offer more advantages. Firstly, they may be cost-effective, as they can drive low-cost technologies and promote competition for cost-reducing RETs. They also set targets that allow more accurate energy planning and policy-making for climate change mitigation. However, experience indicates that quota obligations have not been particularly successful in promoting more costly RETs. Such quotas/ renewables obligations leave a level of uncertainty with regard to the additional costs of deployment. Feed-in tariffs, on the other hand, can provide price certainty to investors and help emerging technologies get off the ground.

Quotas/ renewable portfolio standards offer more advantages over FITs to developing countries.

However, there are also some problems with the use of both kinds of policy incentives. They seldom encourage competition among investors, and they provide insufficient incentives for technological development and innovation. Large subsidies do not encourage developers or manufacturers to reduce costs, although phasing out the FIT could help (Schaeffer and Voogt, 2000; Oliver et al., 2001). Nevertheless, there is considerable potential for selective application and experimentation with these policy incentives in developing countries.

Developing countries need to bear in mind two important aspects when designing integrated innovation policy frameworks using these incentives. First, mobilizing the volume of RETs required for the reduction of energy poverty and climate change mitigation requires an unprecedented level of cooperation between government bodies and the private sector. Therefore enabling such cooperation should be an institutional priority. Second, an analysis of how the more advanced developing countries managed to scale up their capacity for RETs production and use shows that in their integrated frameworks for promotion of RETs they provided dedicated incentives to promote the dual objectives of production on the one hand, and greater deployment and use on the other. Such an approach relies on a greater level of coordination between energy targets and innovation strategies. China, which is placing significant emphasis on RETs, has adopted such an integrated approach, and has emerged as the developing world's most successful developer and installer of RETs in recent times (box 5.11).

Overall success in policy implementation...relies on a greater level of coordination between energy targets and innovation strategies.

Box 5.11: Promoting integrated approaches for increased production and use of RETs

The total capacity of energy generated in China reached 225 GW in 2009. This represents more than a quarter of China's total installed energy capacity. Over the period 2005–2009, wind power in the country increased 30-fold, and it took China less than four years to emerge as the largest supplier of wind turbines, with three Chinese producers among the world's top ten companies by volume of output.

Two main policy instruments seem to have supported the rise of Chinese production capacity: supportive domestic policy targets, and a policy requirement mandating domestic production of wind turbine components. Such aggressive policy targets include the provisional 2020 target to produce over 500 GW of RE capacity, which would include 300 GW of hydro, 150 GW of wind, 30 GW of biomass and 20 GW of solar PV. This represents more than 30 per cent of China's expected installed capacity of 1,600 GW. These targets are underpinned by a number of demand-side policy mechanisms initially set out in the 2005 Renewable Energy Law. These included mandated portfolio standards, feed-in tariffs for biomass, government-regulated prices, concession programmes for wind, and obligations to purchase all new grid-connected renewable power, together with a number of fiscal and R&D support mechanisms.[a] Additionally, the wind turbine industry was subject to a policy requirement of at least 70 per cent domestic content in terms of the value of materials and components. Similarly, domestic incentives have enabled China to become the world's largest producer of solar PV, supplying more than 40 per cent of global output in 2009 (Martinot and Li, 2010).

The rapid growth of the RETs sector has promoted a more comprehensive government policy, with amendments to the Renewable Energy Law in December 2009. These include better coordination and planning of RETs within the overall energy strategy at national and provincial levels, further development of the energy storage policy and smart grids, removing bottlenecks relating to transmission and interconnections, strengthening of requirements for utilities to purchase all RE-generated power, and increases in the levies on electricity sales to meet the increasing volume of RET subsidies.

Source: UNCTAD, based on the Renewable Energy Law of the People's Republic of China (2005).

[a] See: http://www.ccchina.gov.cn/en/NewsInfo.asp?NewsId=5371.

Like China, India too has an effective integrated RET policy. Other developing countries should also consider adopting integrated innovation policy frameworks for RETs that set clear RET targets and promote them for industrial and commercial use. However, not all developing countries will be able to provide an extensive network of financing and capacity-building of the kind found in China or even India, due to institutional and financing constraints that may impede the granting of policy incentives.

Developing countries could consider ways and means to restrict patents on incremental innovations in this field.

Regarding the use of flexibilities under the TRIPS Agreement, all developing countries are hard pressed to promote greater access to knowledge and learning in their domestic contexts. Particularly, given the rising trends in patenting for RETs, as shown in chapter IV, developing countries could consider ways and means to restrict patents on incremental innovations in this field by providing for a high inventive step requirement in their domestic IPR regimes. This has been used in the pharmaceutical sector to prevent innovations involving only minor technical improvements from getting patented.

D. ADOPTION AND USE OF NEW RETs: POLICY OPTIONS AND CHALLENGES

The importance of greater access to technologies needs to be emphasized in international forums such as those dealing with climate change negotiations, as discussed in chapter IV. Access to technologies plays a critical role in the process of accumulation of capabilities in developing countries. However, the experiences of many countries and sectors indicate that lack of easy access to technologies is not the only impediment to developing countries that are seeking to build their technological capabilities. Equally important is the need to improve their technological absorptive capacity, which refers broadly to the tacit elements that facilitate technological learning among firms and enterprises both in the public and private domain within countries.

The lack of technical and human capacity, combined with underdeveloped trade and logistic facilities results in high costs of RETs production that make them uncompetitive on international markets. They also affect the processes involved in RETs adoption and use. In particular, many developing countries lack the technical capability for testing, operation and maintenance of RETs, as noted in chapter III. Creating the requisite technology absorptive capacity of the kind that facilitates the private sector's greater involvement in the development of RETs is critical for the future deployment and scale-up of locally manufactured and adapted technologies, as witnessed in Brazil, China and India. The provision of subsidies for fossil fuels is another important area that hinders the greater use of RETs.

1. Supporting the development of technological absorptive capacity

Fostering the ability to absorb, learn and apply knowledge for the greater use of RETs in countries involves a number of key elements: training of technical support staff to undertake the design, installation and maintenance of RE systems and to interact with users to solve technical problems and provide them with information on equipment operation; supporting engineers, scientists and researchers to enable them to develop new RE systems and processes; educating decision-makers, including economists, administrators, regulators and financial institutions/investors in order to establish better coordination between energy needs and technology choices; and promoting greater public awareness of and consumer confidence in RETs (Benchikh, 2001; Parthan et al., 2010). In this section, capability/competence development is discussed in terms of training, development of adaptation capacities, and education and outreach.

a. Establishing training centres for RETs

Countries would benefit from establishing RET-specific training centres or introducing RET-specific training in established centres domestically.

Countries would benefit from establishing RET-specific training centres or introducing RET-specific training in established centres domestically (similar to the suggestion in chapter IV at the international level). This is because, for the diffusion of RETs to be sustainable, it is necessary to have a well-trained workforce capable of installing, maintaining and adapting RETs, as well as trained target groups such as users, technicians, researchers/scientists, government officials and investors. Such training can take many forms, from formal degrees and certificates to informal workshops and web-based courses. There are a number of examples of training being integral to the sustainable transfer and diffusion of RETs (box 5.12).

Box 5.12: Importance of training for RETs: Experiences of Botswana and Bangladesh

In Botswana, the lack of trained manpower for repair and maintenance of solar energy devices resulted in the failure of those devices, loss of revenue and dwindling consumer confidence in solar technologies (Jain, Lungu and Mogotsi, 2002). To rectify this situation, seven training programmes aimed at progressively increasing skills and expertise were introduced. They included certificate level courses, a national craft certificate programme, a higher diploma for supervisory personnel and a short course for senior managers in decision-making positions.

Similarly, in Bangladesh, investment in training has been central to the success of Grameen Shakti, a non-profit rural enterprise that enables rural communities to lead a better life through the use of RETs. From the start, Grameen Shakti involved the local community in the planning, implementation and maintenance of solar home systems (SHS). It is now planning to involve local people in providing components and servicing in their community, as they would be familiar with the community's needs. To achieve this, Grameen Shakti has started a network of technology centres managed mainly by women engineers, which train other women as solar technicians. At more than 40 technology centres based in rural areas the women undergo an initial 15 days of training on how to assemble and charge controllers and mobile phone chargers, and how to install and maintain solar home systems. Users are also trained to take care of their own systems and to diagnose simple faults. These technology centres are intended to become self-sufficient businesses that will carry out the routine servicing of SHS in return for a fee that will be paid by Grameen Shakti. The centres will be able to take out small loans to purchase tools and equipment and also sell their services directly to customers (Ho, 2010; Barua, 2008).

Source: UNCTAD.

Box 5.13: Renewable energy technologies in Asia: Regional research and dissemination programme

This SIDA-funded capacity-building programme, coordinated by the Asian Institute of Technology, was implemented between 1997 and 2004. It was undertaken within and by national research institutions (NRIs) in six countries: Bangladesh, Cambodia, the Lao People's Democratic Republic, Nepal, the Philippines and Viet Nam. Key components were local training programmes, workshops/seminars, demonstrations and development of training manuals by the NRIs. Target groups were identified for training on specific technologies which focused on operation, installation, trouble-shooting and maintenance of RE systems. Overall, 16 manuals for training courses were published, 46 local seminars/workshops were conducted, 48 courses were completed and 1,100 technicians were trained.

Source: UNCTAD, based on Bhattacharyya and Ussanarassamee (2004).

Several new regional and national donor initiatives are increasingly being designed to address lack of skills and expertise. In South-East Asia, for example a regional training programme funded by the Swedish International Development Agency (SIDA) has had a significant impact (box 5.13).

b. Development of adaptation capabilities

Various systemic failures identified in chapter III necessitate the implementation of domestic policies to support the development of innovative capabilities for adapting and modifying transferred technologies. Supporting R&D and adaptation of RETs through such measures as demonstration projects, dedicated research programmes, specific technology deployment and diffusion activities and development of technologies can reduce perceived investment risks and assist the adaption of technologies to local contexts. Engineering and design (non-R&D) capabilities that enable local firms to experiment with the absorption of technologies are likely to be as relevant as building scientific R&D capabilities in public research institutions.

Engineering and design (non-R&D) capabilities that enable local firms to experiment with the absorption of technologies are likely to be as relevant as building scientific R&D capabilities.

However, imitating the RET innovation systems in developed countries, or in the more advanced developing countries such as China or India, needs to be approached with caution. While there are important lessons for replication in the policy context, specific differences in the socio-cultural context and in economic capacity, the lack of or low level of activities of local enterprises, and low local technological skills may be limiting factors. These limitations imply the need for additional policy support.

Collaboration and joint ventures can be an important means of transferring skills as well as hardware. Other examples of skills transfer include through PPPs or climate technology centres and networks. The latter connect institutions and people around the world working on common themes related to climate change, which also includes learning venues for RETs. One of the lessons drawn from a recent study of existing technology centres and networks by UNEP and Bloomberg (2010) was the need to provide participation incentives, particularly as participating centres cannot make capacity available without compensation and they often cater to favoured vested interests. In addition, such centres should, as far as possible, be located in existing institutions that have the appropriate infrastructure, and their funding needs to be long-term and reliable. Developing countries should actively promote the creation of such centres and networks with the aim of increasing their absorptive capacity specifically for RETs within ongoing work under the climate change agenda.

c. Education, awareness and outreach

Lack of information regarding technologies, user needs, local contexts, and regulations and standards are all barriers to investment in and use of RETs. Education and marketing of RETs at every point along the supply chain – from investors and project developers to users – can help remove some of the barriers. Education should encompass firms, financial institutions, community co-operatives and individuals. Knowledge of the various incentives to invest in and produce energy from RETs, coupled with an awareness of the opportunities small-scale RETs can offer local communities, are all important for stimulating demand as well as supply. Consumer awareness of energy services provided by RETs can further increase demand, thereby providing a positive signal to investors and also public awareness of such services. Policymakers and regulators also need information on how to deal with and integrate RETs into the existing energy system, while project developers need to understand the financial options available and the needs of users.

Improving consumer awareness requires education and outreach of various kinds. This can include advice and/or aid in implementation, the creation of best practice guides, development of comparison and endorsement labels, consultation, dissemination of information and promotional activities (IEA, 2011; Komor and Bazilian, 2005; Oliver et al., 2001). By increasing customers' awareness of the advantages of RETs they would be more likely to agree to pay higher tariffs for "green electricity", and the utilities could guarantee to purchase the corresponding amounts of electricity from RE producers (Ackermann, 2001).

2. Elimination of subsidies for conventional energy sources

Neither the environmental advantages of RETs nor the environmental costs of fossil fuels are currently captured by market mechanisms, which is a very significant problem for policymakers to resolve in order to promote RET-based energy sources. As an initial step, they could eliminate the subsidies for fossil fuels. This may be not be easy because conventional energy technologies tend to enjoy considerable subsidies in many countries, many of which have become embedded in the energy system over time. Targeted and proactive domestic policy interventions could help overcome these challenges and encourage the diffusion of RETs.

a. Removal of subsidies for carbon-intensive fuels

High subsidies for the production and distribution of fossil fuels for power generation can make RETs less competitive than would otherwise be the case. Therefore, their reduction, where possible, should be a key policy objective of governments. There is already a downward trend in subsidies for fossil fuel production, especially coal, in many OECD countries, reflecting the steady privatization and liberalization of energy markets. Many of these countries are switching support from production of electricity generated from fossil sources towards economic restructuring and redeployment of the workforce. A global review of energy subsidies in 2010 measured the shortfall between the costs of supply and the costs to consumers (price-gap approach) in 37 countries (almost all non-OECD countries) that have significant consumption subsidies (IEA, 2010). It found that the consumption subsidies amounted to $312 billion in 2009. Subsidies for the production of fossil fuels (most often offered by OECD countries) have been estimated at another $100 billion per year (GSI and IISD, 2010).

Neither the environmental advantages of RETs nor the environmental costs of fossil fuels are currently captured by market mechanisms…

Germany, where the coal industry had been subsidized for more than 50 years, primarily to support electricity production, offers an example of successful subsidy reform. Total subsidy support reached a peak in 1996, at €6.7 billion, despite declining levels of coal production, but in 2007 such support fell to approximately €2.5 billion, although this still represented an annual support of €90,000 per employee within the industry. The Government has decided that by 2018 all subsidies to the indigenous German coal industry will be phased out (UNEP, 2008).

…which is a significant problem for policymakers to resolve in order to promote RET-based energy sources.

In developing countries, energy subsidies are often considered a tool of social policy, as they protect the poor from the increasingly high prices of fossil fuels. However, this means that many governments pay a disproportionate percentage of their budgetary funds in mitigating the impact of high fuel prices. Moreover, fossil fuel subsidies reduce the incentive to improve efficiency,

Public procurement of renewable energy can provide a strong signal to markets and the private sector about the level of commitments by governments to support long-term targets...

...in addition to providing significant stimulus to technology development and distribution.

and to switch to more reliable and cost-effective forms of energy. They also divert investment away from potential improvements in grid and generation efficiency.

An analysis of fossil fuel subsidy reforms indicates that their removal would result in increases in GDP for both developed and developing countries, ranging from of 0.1 per cent in total for 2010 and 0.7 per cent per year by 2050 (GSI and IISD, 2010). There would also be substantially positive environmental impacts. A recent study projects a 10 per cent reduction in GHG emissions by 2050 if consumer subsidies were to be withdrawn in 20 non-OECD countries (Burniaux et al., 2009). From a social protection perspective, the evidence remains unclear, but it is possible to redirect subsidies towards social protection in a much more targeted manner than is currently being done (IISD, 2010). One potential way could be to specifically limit the subsidies only to the poor in the short and medium term so that they do not bear an undue burden resulting from the removal of subsidies.

b. Carbon and energy taxes

Several countries have successfully introduced carbon-related energy taxes in a bid to improve plant efficiency and reduce emissions. For example, from 1970 to 1990, Sweden invested heavily in RET-related R&D, but without significant deployment of these technologies. It was only with the introduction of carbon taxes in 1991 that substantial progress was made in terms of switching from cheaper electric and oil-fired boilers for district heating to biomass co-generation. As a result of the taxes, the use of biomass increased by more than 400 per cent during the period 1990–2000. This led to a number of follow-on technological developments, such as biomass extraction technologies (Johansson and Turkenburg, 2004). Finland, the Netherlands and Norway are other examples of developed countries that introduced carbon taxes in the 1990s.

The United Kingdom has implemented a tax on energy use for large industrial and commercial customers, known as the Climate Change Levy (CCL). The CCL taxes electricity consumption at 0.456 pence per kWh. The levy encourages voluntary efficiency improvements by raising the price of electricity, but it allows exemptions of up to 80 per cent if participants meet certain efficiency improvement targets. Electricity sourced from renewables is also exempted from the levy. The CCL has been extremely successful in encouraging major energy users to cut their emissions, and it is expected that the instrument will result in at least 5 million tons of CO_2 reductions by 2010.

Tradable emission permits are another widely used policy intervention in industrialized countries (box 5.14).

c. Public procurement of renewable energy

Public procurement of renewable energy can provide a strong signal to markets and the private sector about the level of commitments by governments to support long-term targets, in addition to providing significant stimulus to technology development and distribution. The promotion of RE sourcing, alongside energy efficiency standards and smart networks are part of the ECs Energy 2020 strategy (EC, 2010). Questions have been raised about the potential conflict between procuring higher cost RE and best-value procurement rules, which might lead to concerns over fair competition. Procurement guidelines are important in this respect.

A number of countries have promoted public procurement. For example, the Netherlands, as part of its implementation of the Kyoto Protocol, introduced the Renewable Energy for Public Buildings scheme in 2006, which aims to support a shift to climate-neutral supply of energy for government structures by 2012. During the period 2002–2004, a mandate required that 50 per cent of the consumption of electricity of all government buildings be derived from RE sources.[7] In Sweden, in addition to a pre-existing 1997 Investment Support Programme, the Government estab-

Box 5.14: Tradable emissions permits

Emissions trading schemes have developed as a key policy option to reduce carbon intensity in the electricity sector because of their economic efficiency. Creating liquid carbon markets can help economies identify and realize economical ways to reduce emissions of GHGs and other energy-related pollutants and/or improve efficiency of energy use. The largest tradable permit schemes include the EU's Emission Trading System (EU ETS) and the Kyoto Protocol's Clean Development Mechanism and Joint Implementation mechanism. Other schemes are under development in Australia, New Zealand and the United States.

The EU ETS is the major policy instrument within the EU to reduce GHG emissions. Although some EU member States introduced unilateral energy and carbon taxes, it was decided in 1999 that a cap-and-trade system would be more economically efficient. More than 10,000 sites are currently included in the scheme, representing approximately half of the total CO2 emissions within the EU. Electricity and heat production facilities with a 20 MW capacity or more are a key target group within the scheme.

It has been argued that the electricity sector was the best suited of all sectors to be covered by the EU ETS because it was responsible for one third of the total CO2 emissions in the EU (Svendsen and Vesterdal, 2003). Indeed, many low-cost CO2 emission-reduction opportunities existed within the sector, and companies were relatively well-informed of the opportunities to reduce their CO2 emissions, which would lead to premature trading of emissions. Moreover, the sector was already tightly regulated.

As a result, the power sector had the largest GHG reduction burden under the EU ETS. Allocations were made at a national level, without any overall sectoral target for EU power sector emissions. During the second phase (from 2005 to 2008), the power sector was consistently short on emission allowances and had to purchase them in the market to cover those allowances. This is primarily due to the allocation process at national level, where individual governments have assigned short positions to their electricity producers.

A number of issues have arisen related to the participation of the power sector in the EU ETS. The most important of these is the perception of windfall profits by participating power suppliers that passed along the "costs" (based on market value) of their freely issued allowances to their customers. To counter this, full auctioning of permits to the electricity sector will begin in Phase 3 starting in 2013.[a]

Source: UNCTAD.

[a] For details see Directive 2009/29/EC of the European Parliament and of the council of 23 April 2009, available at: http://eur-lex.europa.eu/LexUriServ/LexUriServ.do?uri=OJ:L:2009:140:0063:0087:EN:PDF.

lished a five-year technology procurement programme in January 1998 specifically for electricity production based on renewables. Total funds for this programme were 100 million Swedish kronor (€11 million). The programme was replaced by the Energy Act, which took effect in 2003. These types of targeted activities are increasingly being supplanted by broader attempts to decarbonize overall energy supply.

3. Applicability of policy incentives to developing countries

This section has discussed two types of incentives to foster the increased use and adaptation of RETs in developing countries. The first type of incentives, intended to enhance the technology absorptive capacity of actors in developing countries, remain very important. Not only do they promote the wider adaptation and use of RETs, they are also the first step in developing incremental innovation capacity across countries. These forms of incentives should be actively promoted through appropriate policy frameworks. The second type of incentives is intended to promote integrated approaches to RETs among those developing countries that have some capabilities for innovation and production of RETs. Such countries should adopt integrated approaches to RETs that simultaneously promote innovation, production and greater adaptation, as discussed in box 5.11. Elimination of subsidies on fossil fuels and sending strong signals that support the use of RETs through public procurement will also be very important. While eliminating subsidies, special safety nets for the poor should be designed to ensure that they are not unduly affected. Measures such as imposition of carbon and energy taxes that are being widely used in industrialized countries will

While eliminating subsidies, special safety nets for the poor should be designed to ensure that they are not unduly affected.

need to be carefully thought through for their socio-economic implications before use in developing- country policy frameworks.

E. MOBILIZING DOMESTIC RESOURCES AND INVESTMENT IN RETs

Particularly in developing countries that face several financial constraints on the introduction and uptake of new technologies...

Financial incentives of various kinds can promote investment in RETs, and facilitate their quicker adaptation and utilization. These incentives need to be developed with an eye on the co-benefits that can be derived from using RETs not only for electricity generation, but also more broadly as a tool for industrial development in countries. All stages of the RETs innovation and adaptation chain require financing, as noted in previous chapters, and will depend on countries' ability to provide a mix of different kinds of financing, including venture capital, equity financing and debt financing. Particularly in developing countries that face several financial constraints on the introduction and uptake of new technologies, governments need to support the private sector in its financing of innovation activities, such as by offering loan guarantees, establishing business development banks or mandating supportive lending by State banks. Governments may also directly fund innovation activities through, for example, grants, low-interest loans, export credit and preferential tax policies (e.g. R&D tax credits, capital consumption allowances). While designing support packages for the private sector, national policymakers need to ensure that technology developers are not overcompensated for their financial and economic risks. However, where it is important to prove the feasibility or viability of a new technology or sector, or where information or coordination failures are being addressed, forms of concessional finance and support may be justified. Some financial incentives that can be provided for the private and public sector are discussed in this section. While designing financial incentives, developing countries should bear in mind the need for financing local enterprises' medium-scale projects that lead to incremental, adaptive learning in RETs, as highlighted in previous chapters of this *TIR*.

...governments need to support the private sector in its financing of innovation activities.

1. Grants and concessional loans

National governments may choose to make funds available at a subnational level to support investments in RE and environmental sustainability. One example is the Swedish Local Investment Program (LIP), which was implemented by the Swedish Environmental Protection Agency. The LIP provides grants to local authorities to make investments in pre-defined areas slated to bring environmental benefits, and it oversees the environmental outcomes. The potential scope for investments is relatively broad, covering energy efficiency as well as renewable power generation projects. The programme was expected to result in a conversion to 2.6 TWh per annum, leading to an annual emission reduction of 1.7 million tons.[8]

Among developing countries, examples include China, which has established a fund in excess of $400 billion to support clean power and RE. Similarly, the Philippines has supported the deployment of renewables through a $2 billion fund. In 2009, the Bangladesh Central Bank established a $29 million fund for similar purposes.

At the regional and international level, several multilateral development agencies contribute to the Clean Technology Fund (CTF) and the Scaling up Renewable Energy programme, both under the Climate Investment Funds.[9] They include subnational grants and concessional finance as core components, and are administered either directly or through national governments.[10]

2. Tendering systems

Tendering systems are quantity-driven mechanisms that aim to promote either investment in RETs or the greater use of RETs for electricity generation. For projects requiring investments, an announcement is made about a proposed investment support package and companies compete on the basis of the amount of capacity they plan to install in the proposed project. For a capacity mechanism, the desired installed capacity is announced, and investors compete on the basis of cost. Both routes use a bidding process in which commercial developers compete to maximize the economic efficiency of RET deployment. This structure was pioneered by the Non-Fossil Fuel Obligation of the United Kingdom, but has been taken up by a number of other countries as well. Tenders are being used in Denmark for offshore wind, in France for wind, biomass and biogas, and in Latvia and Portugal for wind and biomass (Canton and Lindén, 2010).

A number of developing-country governments have used a system of competitive bidding to install fixed quantities of RE capacity. For example, China initiated a wind power concession between 2003 and 2007, part of which involved a competitive bidding process for additional capacity on a yearly basis. A total of 3.4 GW was added, but the scheme was later substituted by feed-in tariffs for new capacity. In Brazil, the PROFINA programme sought tenders for 3.3 GW capacity using small hydro, wind and biomass under its first phase in 2009. Other countries in Latin America have implemented similar tendering auctions, with Uruguay offering 60 MW of wind, biomass and small hydro in 2009, and Argentina offering 1 GW. Peru has indicated its willingness to tender up to 500 MW of renewable capacity by 2012 (REN21, 2010).

3. Fiscal measures

Fiscal measures relate to taxes and expenditures, and have been used extensively to support the deployment of renewables-sourced electricity generation. These may be in the form of tax exemptions, reductions or credits.

One of the earliest tax credit programmes was the Japan Solar Roof programme, which led to Japan becoming the world's leading installer and manufacturer of grid-connected PV systems in 2005. In India, the government has reduced its customs levy on imports of machinery, instruments, equipment and appliances used in solar PV and solar thermal plants to 5 per cent. In Africa, Kenya and Zimbabwe have recently removed excise tax on PV systems.

Since 2003, China has operated preferential tax policies for RE. Foreign investment in biogas and wind projects have benefited from reduced income tax rates of 15 per cent. There are a number of RE enterprises and development projects that are eligible for tax reductions. Wind turbines, solar PV modules and their components also benefit from preferential excise rates. In addition, China provides significant capital subsidies for solar PV installations. For large-scale building systems (50 kw+), and for land-based grid-connected systems (300 kw+) it offers capital investment subsidies of up to 50 per cent, and for off-grid projects the capital subsidies can reach 70 per cent. Subsidies are also helping to support a solar PV pipeline of up to 500 MW, which is expected to be completed by 2012. In other parts of Asia, Indonesia introduced a 5 per cent tax credit for RETs in 2010, and the Philippines has granted tax exemptions and removed VAT for investments in RETs. In India, there are specific tax exemptions and accelerated depreciation allowances for investments in wind power.

Fiscal measures relate to taxes and expenditures, and have been used extensively to support the deployment of renewables-sourced electricity generation.

Mexico introduced the Accelerated Depreciation for Environmental Investment programme in 2005, which allows a deduction of up to 100 per cent of the investment in the first year of an RE-related investment. The plant has to remain operational for at least five years and should be able to produce a stipulated minimum output in compliance with local legislation.[11]

4. Facilitating foreign direct investment in RETs

Facilitation of foreign direct investment (FDI) involves creating enabling conditions for attracting investors, and this will remain a key goal of the broader innovation policy framework for RETs. Investors, both foreign and domestic, consider a number of factors when making investment decisions, particularly those concerning the domestic environment for investment. They assess how risky or difficult it will be to make an investment in a given country using a given technology, and add this to the expected costs. Broadly, investors look for political and macroeconomic stability, an educated workforce, adequate infrastructure (transportation, communications and energy), a functioning bureaucracy, the rule of law, a strong financial sector, and ready markets for their products and services. Policy barriers differ fundamentally from country to country, and from sector to sector. In the case of RETs, there are many factors that shape national energy policies, including history, politics, geography and chance, on the one hand, and innovation and production climate on the other. Various studies have noted that many developing countries, particularly the least developed among them, are not getting their full share of investment in renewables because their existing policies make them unattractive, except for projects with the highest potential returns (Amin, 2000; Chandler and Gwin, 2008; Point Carbon, 2007; Dayo, 2008; Neuhoff, 2008; Cosbey and Savage, 2010).

Countries should consider ways and means to make sure FDI leads to building technological absorptive capacity in RETs.

Government-led efforts to facilitate the uptake of RETs need to identify and overcome the main barriers to trade and investment in these technologies. FDI could be an important source of technology for recipient countries, and could also lead to the accumulation of tacit know-how in the host countries with regard to plant operation, maintenance and RETs use. Countries could attract FDI through policy incentives, but at the same time they should consider ways and means to make sure FDI leads to building technological absorptive capacity, such as requiring FDI to include training and sharing of know-how on production, maintenance and use of RETs.

Box 5.15 presents some recent global trends in RE-related FDI.

Box 5.15: FDI in the global market for solar, wind and biomass

FDI flows to the renewable energy industry have increased significantly in recent years. In 2003, the estimated value of such flows was less than $10 billion, and rose steadily thereafter, reaching close to $121 billion in 2008. The global financial crisis in 2009 interrupted this trend, resulting in a fall in RE-related FDI to below $100 billion. Continued economic weakness and tight credit conditions hampered investments in 2010, with overall FDI dropping further, to roughly $61 billion. However, investments in RET manufacturing projects experienced a rebound in 2010, rising to $20.2 billion from $13.6 billion in 2009.

Box figure 5.15.1 Evolution of FDI flows in RE-based electricity generation and RETs manufacturing, 2003–2010 ($ billion)

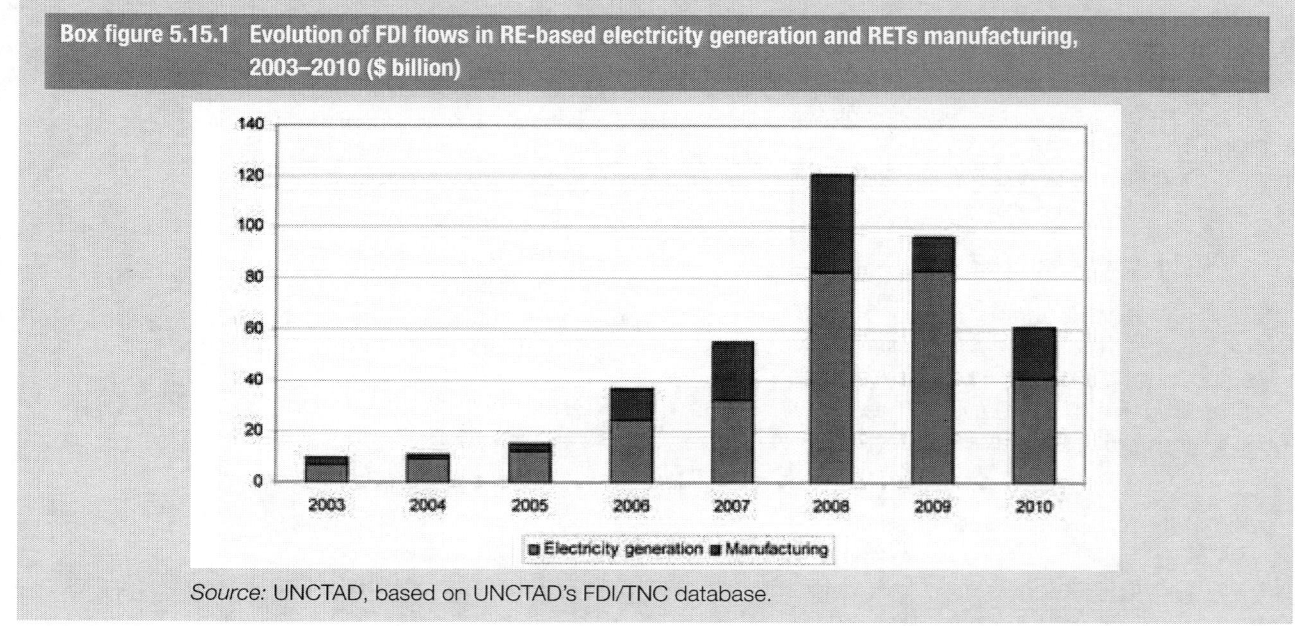

Source: UNCTAD, based on UNCTAD's FDI/TNC database.

Box 5.15: FDI in the global market for solar, wind and biomass (continued)

FDI in the renewable energy market can flow to two main segments: (i) RE-based electricity generation, also referred to as the downstream segment; and (ii) manufacturing of RETs, also referred to as the upstream segment. The latter includes the fabrication of equipment needed in the downstream segment, such as wind turbines, solar panels, or plants and equipment needed in the production of biofuels.

During the period 2003–2010, most of the merger and acquisition activity took place in the market for RE-based electricity generation. In terms of the number of announced greenfield investment projects, there was a more balanced distribution between the upstream and downstream markets, with 46 per cent and 54 per cent respectively. During the period 2003–2010, UNCTAD estimates that FDI in RE projects amounted to $406 billion. The regional distribution of these investments is shown in the box table below.

Box table 5.15.1 Shares of FDI in renewable energy technologies by investing and host economic groupings, 2003–2010 (Per cent)

	Reporting (investing) regions			
Partner (host) region	Developed economies	Developing economies	South-East Europe and CIS	Total
Developed economies	55.2	1.7	0.1	57.0
Developing economies	31.3	7.5	0.5	39.3
South-East Europe and CIS	3.0	0.4	0.4	3.7
Total	89.5	9.6	0.9	100

Source: UNCTAD, based on UNCTAD's FDI/TNC database

Developed economies have been by far the biggest investors in the REs industry. Over the period 2003–2010, they accounted for nearly 90 per cent of the reported FDI flows, followed by developing economies with 9.6 per cent and South- East Europe and the Commonwealth of Independent States (CIS) with 0.9 per cent. In terms of destination of these flows, developed economies have been the main target, attracting 57 per cent of total FDI, whereas developing countries attracted 39.3 per cent and South-East Europe and the CIS 3.7 per cent. Although developed economies stand out as the main recipients of FDI in RE projects, developing economies are becoming increasingly popular investment destinations in the manufacturing of environmental technology products. Indeed, during the period 2007–2010 developing economies as a whole attracted more investment projects in this area, by value, than developed countries. However, this generalization obscures the fact that China accounted for a large share of this investment.

Source: UNCTAD.

Two important caveats need to be noted with respect to FDI. First, removing some or all of the general barriers to investment may not necessarily lead to greater FDI for RETs in developing countries. Second, not all investment is desirable investment. In the rush to attract FDI in the RETs sector, governments should not abandon their other public policy objectives, or waive the due diligence process in screening and regulating investors in RETs projects that they would apply to investors in other sectors.

F. NATIONAL POLICIES ON SOUTH-SOUTH COLLABORATION AND REGIONAL INTEGRATION OF RETs

South-South collaboration presents new opportunities for increasing the use and deployment of RETs not only through trade and investment channels, but also through technology cooperation, and this can be facilitated by governments, intergovernmental organizations and/or development banks (as noted in chapter IV). Such cooperation can also be mediated by private sector owners of RETs, although this is less frequent. Technology cooperation can take several forms, ranging from training foreign nationals in the use and maintenance of RETs to supporting research in partner countries to adapt existing technologies to local needs. It can also include outright grants of RET-related intellectual property rights, or licensing on concessionary terms. In several cases a developed-country institution has been involved in bringing developing-country partners together for this sort of cooperation, as discussed below. The benefits of such collaboration could include the wide dissemination of RETs among developing countries, along with the commensurate benefits discussed earlier in this analysis.

1. Best practices in trade and investment

There are several modes by which trade and investment can help in the South-South dissemination of RETs. They include the simple flow of traded goods, the flow of investments or joint investments in new RETs (such as energy generation facilities), or the award of a concession for anything from the building of new capacity to a complete turnkey operation that involves building, operating and maintaining new RE capacity. As a general rule, the greater the involvement of the more technologically advanced developing country in the venture, the greater will be the degree of technology learning in the recipient country, which will provide a better basis for building domestic capabilities for production in that sector. However, the extent of technology transfer and learning will vary depending on the characteristics of different RETs.

Trade in RETs and associated goods is one way to effect technology transfer, though by itself it offers a limited means of disseminating technologies. Ideally, it would be accompanied by ancillary investments and training where the technologies are novel. It may also be limited in generating backward and forward linkages and cultivating domestic capabilities in the sector. Brazil, for example, exports hydrous ethanol to facilities in Jamaica and Trinidad that then dehydrate the stock to create anhydrous ethanol (USDA, 2008). The finished product is exported tariff-free to the United States under the Caribbean Basin Initiative free trade agreement, thereby avoiding a 54 cents/litre tariff to which Brazilian exports would have been subject. This example shows that trade alone does not necessarily lead to the dissemination of RETs, since there is no uptake of the ethanol as fuel in the processing countries.

Private sector investment, a common form of South-South collaboration in general, is an effective means of disseminating RETs.

Another example of South-South collaboration is that of China's Trina Solar, which was contracted to export its PV modules used by an Indian company (Lanco Infratech) to build a 5 MW solar PV plant in India's state of Rajasthan – the country's largest at the time of building. Here the source firm was simply part of the developer's global supply chain, which was beneficial for both firms in terms of increased learning and economic viability.

Often, host countries seek to develop their RET capacity by means of concessions to build, and potentially operate and maintain, RE facilities. For example India's Suzlon Energy reported in February 2011 that it had won a bid to build, operate and maintain a 218 MW wind power installation in the states of Ceará and Rio Grande do Norte in Brazil. The ultimate impact of such deals will probably depend on the details of the arrangement (which were unavailable in this case). Purchasing governments can specify various requirements such as local content, the nature of the joint venture, the use of a specified percentage of nationals in management, use of local R&D, and other requirements designed to foster greater backward and forward linkages, and to stimulate domestic industrial capacity. However, in many instances, performance requirements contravene WTO rules and regional trade agreements.[12] In this context, export-related performance requirements are effective at creating linkages and spillover effects within the host economy, while those related to technology sharing and joint ventures are on the whole less effective (Moran, 1999 and 2001; UNCTAD, 2003; Kumar, 2005).

Private sector investment, a common form of South-South collaboration in general, is an effective means of disseminating RETs. For example, in 2010 China's Suntech Power Holdings signed an agreement to develop solar PV capacity of up to 100 MW – an investment estimated to be worth $350–$400 million (Reuters, 2010). The investment was to involve a South African company as a local partner. In 2007, Suzlon Energy established a factory in Tianjin, China, to manufacture rotor blades, generators, hubs and other wind turbine components, which would be able to produce a capacity equivalent of 600 MW per annum, and it has also established an on-site

R&D centre. Suzlon has also partnered with domestic investors on Chinese wind installations, and has recently opened a sales office in the Republic of Korea on the expectation of strong future sales in that country.

As with concession agreements, these sorts of investments vary in their impact according to the specifics of each case, with government policy being a key determinant. As noted earlier in this Report, another key determinant is domestic capacity to absorb the technology in question, and to capture spillover benefits. Use of local qualified staff in management and senior operating capacities creates potential for learning by doing in aspects that include the technology itself, its operating procedures and management practices. In addition, the availability of local firms and organizations with the capacity to partner with or act as suppliers to the investors stimulates domestic capacity by exposing those firms to globally competitive standards and practices.

Developing countries need to be able to exploit the immense potential of South-South investment in RETs through institutional incentives, while at the same time ensuring that those investments promote national strategic interests and development goals. For example, Abellon Clean Energy of India recently announced a plan to invest in Ghana, in an operation that would involve the cultivation of feedstocks and the refining of solid biofuels for export. The plan envisages cultivation on 10,000 hectares of "derelict" land (Bloomberg, 2011). This investment could represent an unqualified boon for Ghana: company projections are for employment of 21,000 farmers and 4,000 factory workers. But it will be important for Ghana to ensure that this investment aligns well with domestic public policy goals by ensuring, among other things, that the project will not unduly exacerbate existing water shortages through irrigation or industrial use, that industrial effluent will not exceed acceptable levels, and that the legal status of the land in question is clearly defined.

Developing countries need to be able to exploit the immense potential of South-South investment in RETs through institutional incentives.

Newer projects for regional cooperation are in the process of being implemented in countries, some of which are regional in nature, and they offer hope that regional ventures may well provide significant solutions in this area. A promising venture is the Turkana Wind Power project currently under way in East Africa (box 5.16). The projected capacity of the Turkana wind corridor is 300 MW, which would be twice the current installed capacity of Chile (discussed in box 5.3 above).

2. Best practices in technology cooperation

Cooperation agreements are another form of South-South collaboration on RETs, which involve sharing of technologies, training or joint R&D between countries and firms. These may be led by the governments involved, and/or facilitated by intergovernmental organizations, or multilateral or regional development banks. They may also be led by the private sector, but usually the private sector participates as a partner in efforts organized by others.

Cooperation agreements are another form of South-South collaboration on RETs, which involve sharing of technologies, training or joint R&D between countries and firms.

There are a number of examples of such arrangements. Under the Bilateral Memorandum of Understanding to Advance Cooperation on Biofuels between Brazil and

Box 5.16: The Turkana Wind Power Project, Kenya

The Lake Turkana Wind Power Project located near Lake Turkana in north-western Kenya aims to take advantage of the winds that are channeled through the Turkana Corridor between the Ethiopian and Kenyan highlands. The project envisages the construction of a wind park comprising 353 wind turbines, each with a capacity of 850 KW. The initial phase of this wind farm was supposed to start production in June 2011, and it is expected that when it reaches full production by July 2012 it will provide 300 MW of clean power to the country's national grid, which amounts to 30 per cent or more that would be added to the total existing installed capacity in Kenya.

Source: UNCTAD, based on Lake Turkana Wind Power Project website at: http://laketurkanawindpower.com/default.asp

the United States in 2007 the Organization of American States, as an institutional partner to the agreement, promotes technical assistance and policy reform in a number of countries (the Dominican Republic, El Salvador, Haiti, Guatemala, Jamaica, and Saint Kitts and Nevis) to enable the development and local use of ethanol, based in part on Brazilian experience and expertise. The Brazilian industry association for ethanol producers (UNICA) is involved as a partner.

Developing countries may face a variety of constraints in each of these areas, but there are also several opportunities.

In China, UNIDO helped create the Gansu Natural Energy Research Institute, the International Solar Energy Center for Technology Promotion and Transfer and the Asia-Pacific Research and Training Center for Solar Energy. The two former institutions cooperate with UNIDO to conduct training and convene conferences for foreign nationals in solar and other RETs. Between 1991 and 2008, they trained more than 800 solar energy users from 104 countries, most of them from other developing countries. They routinely send staff to the field to conduct training and provide technical assistance in developing countries. This is an excellent model of capacity-building coupled with R&D and supported by intergovernmental agencies.

Another example of technology cooperation is Brazil's willingness to share its expertise in the production of ethanol. Largely through its agricultural research institute, EMBRAPA, the Brazilian Government has carried out extension work in a number of developing countries, including 15 African countries that plan to produce ethanol with technology and supervision provided by Brazil.[13] In Ghana, the Brazilian firm Constran has invested more than $300 million in developing the infrastructure necessary to produce up to 180 million litres of ethanol annually (*El Pais*, 2008). EMBRAPA has set up a permanent facility in Ghana to facilitate the ongoing projects there, which includes supporting domestic R&D. This model involves private sector and government cooperation in seeking economic expansion and investment opportunities, and in the process building capacity in the host countries. Several projects mentioned in this chapter are being implemented in a number of developing countries. For example, the the Barefoot College and the Lighting a billion Lives Initiative (boxes 5.1 and 5.2) initiated in India are now being implemented in several other developing countries and LDCs.

G. SUMMARY

This chapter has presented elements of an integrated innovation policy framework for RETs use, adaptation, innovation and production in developing countries. The concept of such a framework envisages linkages between two very important and complementary policy regimes: national innovation systems that provide the necessary conditions for RETs development, on the one hand, and energy policies that promote the gradual integration of RETs into industrial development strategies on the other. The chapter suggests that such a framework is essential for harnessing benefits from the virtuous cycle of interaction between RETs and science, technology and innovation.

The analysis has identified elements of a national integrated innovation framework for developing countries under the following five headings:

(i) setting policy strategies and goals;
(ii) providing policy incentives for R&D, innovation and production of RETs;
(iii) providing policy incentives for greater technological absorptive capacity that is needed for adaptation and use of available RETs;
(iv) promoting domestic resource mobilization for RETs in national contexts; and finally,
(v) exploring newer means of improving innovation capacity in RETs, including South-South collaboration.

Developing countries may face a variety of constraints in each of these areas, but there are also several opportunities. The chapter has presented numerous examples of initiatives, including policy incentives, which have worked well in both developed and developing countries. Some important lessons from these initiatives are summarized below.

First, the success of a number of emerging economies in developing technological capabilities over time is largely attributable to the role of national governments in providing strategic, targeted policy support for the use, adaptation and deployment of RETs. However, the kinds of incentives, which have been used by industrialized countries or by the larger developing countries such as India or China may not be applicable to other developing countries due to their less favourable circumstances. The chapter also highlights some of the policy incentives that need to be approached with caution. Of special note are those related to carbon taxes, which may not be relevant or useful in the context of many developing countries.

Second, developing countries should consider diversified energy regimes that give priority to the deployment of REs most suited to their contexts, while ensuring that conventional energy sources are not subsidized extensively. This includes rectifying unfavourable arrangements such as monopoly providers of power that control both generation and distribution, and where there are no requirements (or no requirements on reasonable terms) for those providers to purchase independently produced power. Some means of resolving this issue are through the provision of fiscal incentives for renewables (e.g. feed-in tariffs), or mandated requirements for sourcing electricity generation from a given percentage of renewables (e.g. renewable purchase obligations).

Third, success in eliminating, or at least reducing, energy poverty through the use of RETs does not necessarily require large-scale projects with huge investments. Smaller initiatives have been highly successful as off-grid solutions to rural electricity, and offer considerable potential for replication.

Success in eliminating, or at least reducing, energy poverty through the use of RETs does not necessarily require large-scale projects with huge investments.

Fourth, creating an integrated innovation policy framework of the kind outlined in this chapter should not be viewed as a daunting exercise. Nor is a comprehensive policy framework a prerequisite for beginning to harness the potential of RETs for energy access and sustainable development. In the developing-country context, a few well-coordinated incentives can go a long way in achieving significant results, and these can serve as the building blocks for an integrated framework in the years to come. Further, many countries may already be providing several of the policy incentives listed here. The emphasis in such cases needs to be on enhanced coordination to reach targets in RETs use, promotion and innovation.

Fifth, countries will need to experiment with different policy combinations, and this learning process could have positive impacts on the RETs sector. With time, incentives and policy frameworks evolve in tandem with the technological sophistication of the sector.

Countries will need to experiment with different policy combinations, and this learning process could have positive impacts on the RETs sector.

Finally, as noted in chapter IV, and stressed elsewhere in this *TIR*, developing countries will need the support of the international community to benefit from the potential that RETs offer for alleviating (and eventually eliminating) energy poverty and for climate change mitigation. Forging a strong partnership with the international community could lead to the widespread dissemination of environmentally sustainable technologies worldwide, resulting in enhanced economic development and greater opportunities for large segments of populations that have been left behind in the process of globalization.

NOTES

[1] http://ec.europa.eu/energy/renewables/targets_en.htm

[2] See REN21 (2010) for a list of renewable electricity production targets set by different developing countries as of 2010.

[3] Couture and Gagnon (2010) present seven ways to structure a FIT. For country case studies, see, for example, Chua, Oh and Goh (2011) for Malaysia, del Río González (2008) for Spain and Schaeffer and Voogt (2000) for Denmark, Germany and Spain.

[4] Article 27 of the TRIPS Agreement specifies that 'novelty', 'industrial applicability/utility' and 'inventive step' are the criteria for grant of patents, the provision does not define these criteria. As a result, they can be interpreted in different ways within national regimes.

[5] Brazil, China and India have advocated stronger use of TRIPS flexibilities at the UNFCCC intergovernmental meetings, including the greater use of compulsory licences.

[6] Technological absorptive capacity is a critical prerequisite for countries across many sectors (as discussed in TIR 2010 in the context of agricultural innovation).

[7] http://www.rijksoverheid.nl/ministeries/ienm.

[8] See: www.naturvardsverket.se/en/In-English/Menu/Legislation-and-other-policy-instruments/Economic-instruments/Investment-Programmes/Local-Investment-Programmes-LIP/.

[9] CIF are channelled through the African Development Bank, the ADB, the European Bank for Reconstruction and Development, the Inter-American Development Bank and the World Bank Group.

[10] The CIF is composed of the Clean Technology Fund (CTF) and the Strategic Climate Fund (SCF). Each of them is governed by a separate Trust Fund Committee having equal representation from contributor and recipient countries (see: http://www.climateinvestmentfunds.org/cif/sites/climateinvestmentfunds.org/files/Financing%20Modalities%20nov2010_110810_key_document.pdf).

[11] www.ine.gob.mx.

[12] The WTO Agreement on Trade-Related Investment Measures prohibits performance requirements related to exports and imports, as well as domestic content requirements, if they are used as a condition for obtaining some advantage. The WTO's Agreement on Subsidies and Countervailing Measures (Article 3) similarly prohibits subsidies that are conditional on export performance or the use of domestic content.

[13] The 15 countries are: Benin, Burkina Faso, Cape Verde, Gambia, Ghana, Guinea, Guinea-Bissau, Côte d'Ivoire, Liberia, Mali, Niger, Nigeria, Senegal, Sierra Leone and Togo.

REFERENCES

Ackermann T (2001). Overview of government and market driven programs for the promotion of renewable power generation. *Renewable Energy*, 22(1-3): 197–204.

Almendras JR (2010). Philippine energy market outlook and the national renewable energy plan. Presented at Renewable Energy and Conference Expo Manila 2010. Manila, Philippines, 2 December 2010. Available at: http://www.doe.gov.ph/Sec%20Corner/Philippine%20Energy%20Market%20Outlook%20and%20The%20 National%20RE%20Plan.pdf.

Amin A-L (2000). The power of networks: Renewable electricity in India and South Africa. Brighton, University of Sussex.

Barua D (2008). Rapidly growing solar installer provides clean cooking as well. The Ashden Awards for Sustainable Energy case study. Available at: http://www.ashdenawards.org/files/reports/grameen_case_study_20081105.pdf.

Benchikh O (2001). Global renewable energy education and training programme (GREET Programme). *Desalination*, 141(2): 209–221.

Bhattacharyya SC and Ussanarassamee A (2004). Decomposition of energy and CO_2 intensities of Thai industry between 1981 and 2000. *Energy Economics*, 26(5): 765–781.

Bloomberg (2011). India's Abellon plans 25,000 biofuel jobs in Ghana, UNDP says. 24 February. Available at: http://www.bloomberg.com/news/2011-02-24/india-s-abellon-plans-to-create-25-000-biofuel-jobs-in-ghana-undp-says.html (accessed 24 February).

Burniaux J-M, Château J, Dellink R, Duval R and Jamet S (2009). The economics of climate change mitigation: How to build the necessary global action in a cost-effective manner. Paris, OECD Publishing, June.

Canton J and Lindén A J (2010). Support schemes for renewable electricity in the EU. Brussels, European Commission, Directorate-General for Economic and Financial Affairs.

Chandler W and Gwin H (2008). Financing energy efficiency in China. Washington, DC, Carnegie Endowment for International Peace.

CNE (2009). Modelación del recurso solar y eólico en el norte de chile. Comisión Nacional de Energia. Available at: http://www.cne.cl/cnewww/export/sites/default/05_Public_Estudios/descargas/ModelacionRecursoSolarEolico.pdf.

Chua SC, Oh TH and Goh WW (2011). Feed-in tariff outlook in Malaysia. *Renewable and Sustainable Energy Reviews*, 15(1): 705–712.

Cleantech Open (2010). Driving innovation trends, technology and investment. Preliminary report. Redwood City, CA, November 2010.

Couture T and Gagnon Y (2010). An analysis of feed-in tariff remuneration models: Implications for renewable energy investment. *Energy Policy*, 38(2): 955–965.

Cosbey A and M. Savage (2010). RETs for sustainable development: Designing an integrated response for development. Background paper for TIR 2011, UNCTAD, Geneva.

Curren J (2010). The design of adapted feed-in tariff for African countries. Presented at REEEP-SERN Regional Workshop on Feed-In Tariffs. Johannesburg, November 2010.

Dayo FB (2008). Clean energy investment in Nigeria: The domestic context. Manitoba, International Institute for Sustainable Development.

del Río González P (2008). Ten years of renewable electricity policies in Spain: An analysis of successive feed-in tariff reforms. *Energy Policy*, 36(8): 2917–2929.

El Pais (2008). El desembarco africano de Brasil, Río de Janeiro. Available at: http://www.elpais.com/articulo/internacional/desembarco/africano/Brasil/elpepuint/20081125elpepuint_22/Tes (accessed 24 February 2011).

EC (2010). Energy 2020: A strategy for competitive, sustainable and secure energy. Communication from the Commission to the European Parliament, The Council, The European Economic and Social Committee and the Committee of Regions. Brussels.

Emerging Energy Research (2011). Brazil leads Latin america wind energy markets to 46 GW by 2025. Available at: http://www.emerging-energy.com/content/press-details/Brazil-Leads-Latin-America-Wind-Energy-Markets-to-46-GW-by-2025/29.aspx (accessed 19 September 2011).

Government of the Philippines (2008). Renewable Energy Act of 2008. Manila. Available at: http://www.lawphil.net/statutes/repacts/ra2008/ra_9513_2008.html (accessed 19 September, 2011).

GSI (Global Subsidies Initiative) and IISD (International Institute for Sustainable Development) (2010). Relative subsidies to energy sources: GSI estimates. Available at: www.globalsubsidies.org/files/assets/relative_energy_subsidies.pdf.

Ho M-W (2010). Grameen Shakti for renewable energies. London, Institute for Science in Society. Available at: http://www.i-sis.org.uk/grameenShaktiRenewableEnergies.php (accessed 15 August 2011).

IEA (2010). *World Energy Outlook 2010*. Paris, OECD/IEA.

IEA (2011). *Clean Energy Progress Report*. Paris, OECD/IEA.

IISD (International Institute for Sustainable Development) (2010). The effects of fossil-fuel subsidy reform: A review of modelling and empirical studies. Winnipeg.

Jara W (2007). Energía eólica: la experiencia de endesa chile. Presented at Seminario Oportunidades en el Desarrollo de las Redes de Energía Electrica: Conductores de Alta Capacidad y Parques Eólicos, Chile, 2007. Available at: http://www.cigre.cl/sem_9_sept/presentaciones/endesa%20eco.pdf.

Jain PK, Lungu EM and Mogotsi B (2002). Renewable energy education in Botswana: Needs, status and proposed training programs. *Renewable Energy*, 25(1): 115–129.

Johansson TB and Turkenburg W (2004). Policies for renewable energy in the European Union and its member states: An overview. *Energy for Sustainable Development*, 8(1): 5–24.

Komor P and Bazilian M (2005). Renewable energy policy goals, programs, and technologies. *Energy Policy*, 33(14): 1873–1881.

Kumar N (2005). Performance requirements as tools of development policy: Lessons from experiences of developed and developing countries for the WTO Agenda on trade and investment. In: Gallagher K, ed. *Putting Development First*. London, Zed Press.

LAWEA (2010). 2009-2010 energía eólica en América Latina. Wind Energy Association for Latin America. Available at: http://www.lawea.org/YearBook/2009-2010/EspanolFinal/espanol.pdf.

Martinot E and Li J (2010). China's latest leap: An update on renewables policy. Available at: http://www.renewableenergyworld.com/rea/news/print/article/2010/07/renewable-energy-policy-update-for-china.

Moner-Girona M (2009). A new tailored scheme for the support of renewable energies in developing countries. Available at: http://www.sciencedirect.com/science/article/pii/S0301421508007222.

Moran TH (1999). Foreign direct investment and development: A reassessment of the evidence and policy implications. In: OECD. *Foreign Direct Investment, Development and Corporate Responsibility*. Paris, OECD.

Moran TH (2001). Parental supervision: The new paradigm for foreign direct investment and development. Washington, DC, Peterson Institute for International Economics.

Negro SO, Hekkert MP and Smits R (2008). Stimulating renewable energy technologies by innovation policy. Utrecht, Department of Innovation Studies, Utrecht University.

Neuhoff K (2008). International support for domestic climate policies: Policy summary. Cambridge, Climate Strategies. Available at: http://www.climatestrategies.org/research/our-reports/category/40/99.html.

Oliver T, Lew D, Redlinger R and Prijyanonda C (2001). Global energy efficiency and renewable energy policy options and initiatives. *Energy for Sustainable Development*, 5(2): 15–25. June.

Parthan B, Osterkorn M, Kennedy M, Hoskyns St. J, Bazilian M and Monga P (2010). Lessons for low-carbon energy transition: Experience from the Renewable Energy and Energy Efficiency Partnership (REEEP). Energy for Sustainable Development, 14(2): 83–93.

Point Carbon (2007). Carbon 2007: A new climate for carbon trading. Available at: http://www.pointcarbon.com/polopoly_fs/1.189!Carbon_2007_final.pdf.

REN21 (2009). *Renewables Global Status Report 2009*, update. Paris, REN21 Secretariat.

REN21 (2010). *Renewables 2010 Global Status Report 2010*. Paris, REN21 Secretariat.

Reuters (2010). Suntech signs MOU to build solar power plants in S. Africa, 26 August. Available at: http://af.reuters.com/article/topNews/idAFJOE67P00J20100826.

Schaeffer GJ and Voogt MH (2000). Policy options for the stimulation of renewable energy in a liberalising market: An analysis. In: World Renewable Energy Congress VI. Oxford, Pergamon: 1685–1688.

Scottish Executive (2009). The Renewables Obligation Order 2009. Scotland. Available at: http://www.legislation.gov.uk/ssi/2009/140/pdfs/ssien_20090140_en.pdf.

Sovacool B (2010). A comparative analysis of renewable electricity support mechanisms for Southeast Asia. *Energy Policy*, 35(4), April: 1779–1793.

Svendsen GT and Vesterdal M (2003). How to design greenhouse gas trading in the EU? *Energy Policy*, 31(14): 1531–1539.

UNCTAD (2003). *Foreign Direct Investment and Performance Requirements: New Evidence from Selected Countries*. United Nations publication, sales no. E.03.II.D.32. New York, United Nations.

UNCTAD (2010). *Technology and Innovation Report 2010: Enhancing Food Security in Africa through Science, Technology and Innovation*. New York and Geneva, United Nations.

USDA (2008). Brazil Bio-Fuels Annual – Ethanol 2008, no. BR8013. Washington, DC.

UNEP (2008). *Reforming Energy Subsidies: Opportunities to Contribute to the Climate Change Agenda* (No. 1). Available at: http://www.unep.org/pdf/PressReleases/Reforming_Energy_Subsidies.pdf

UNEP and Bloomberg (2010). *Global Trends in Sustainable Energy Investment 2010: Analysis of Trends and Issues in the Financing of Renewable Energy and Energy Efficiency.* United Nations Environment Programme. Available at: http://bnef.com/Download/UserFiles_File_WhitePapers/sefi_unep_global_trends_2010.pdf.

UN/DESA (2009). A global green new deal for climate, energy, and development. Available at: http://www.un.org/esa/dsd/resources/res_pdfs/publications/sdt_cc/cc_global_green_new_deal.pdf.

van Alphen K, Kunz HS and Hekkert MP (2008). Policy measures to promote the widespread utilization of renewable energy technologies for electricity generation in the Maldives. *Renewable and Sustainable Energy Reviews*, 12(7), September: 1959–1973.

ZEP (2008). EU demonstration program for CO_2 capture and storage (CCS). European Technology Platform for Zero Emission Fossil Fuel Power Plants (ZEP), October.

CONCLUSION

CHAPTER VI

CONCLUSION

Energy poverty remains the key issue in the interface of climate change and development. As highlighted in this *TIR,* RETs offer a distinct possibility of tackling the dual challenge of climate change and energy poverty. Established RETs offer an important means not only of reducing energy poverty, but also of complementing national strategies for climate change mitigation. All countries have the potential to transition to an energy sector that contains suitable mixes of conventional and renewable energy sources in order to alleviate energy poverty, and to reposition their economies on greener catch up trajectories. However, the challenge for developing countries and LDCs is to build their technological capabilities and domestic markets so that they can promote the use, large-scale adaptation, production and innovation of RETs for development of manufacturing and other sectors of the economy. The ensuing added benefits in terms of job creation and export potential will lend the much-needed impetus to their economic development.

These are not individual challenges, as much of the analysis in this Report points out. Like many other issues in development today, there is an increasing convergence of causative factors. The following areas of interface stand out in the debates on technology, innovation and climate change that are of particular relevance for all developing countries.

(i) Developing countries are at a cross roads currently with regard to exploring the extent to which already established RETs can help alleviate energy poverty by complementing traditional energy sources. Access to energy is such a crucial element in contributing to well-being and development that it is repeatedly being referred to in policy debates as the "missing MDG".

(ii) The structural transformation of countries is fundamental to development, and this relies strongly on the growth of national technological capabilities. Deployment of RETs can be a valuable part of an overall industrialization effort. However, the lack of reliable power supplies is a crippling bottleneck in the process of industrialization in developing countries and LDCs. In particular, it inhibits growth of productive sectors in many of these countries. It also impedes the development and performance of other sectors that are potentially important to the process of industrialization and development, such as services, tourism, agricultural processing, and others that depend on a reliable, high quality power supply regime. This Report has therefore stressed the need for recognizing that energy security and technological capabilities have a virtuous relationship. Energy security is key to the provision of the necessary physical infrastructure that promotes high levels of enterprise growth in the early stages of structural change, and technological capabilities are a fundamental prerequisite for the increased adaptation and use of RETs within domestic economies. National strategies for sustainable development that include promoting a greater use of RETs are likely to be constantly undermined by the lack of technological and innovation capacity which are required not only for R&D and innovation of new RETs, but also for adaptation, dissemination and use of existing RETs.

(iii) Policy developments at the international level tend to have a significant impact on national aspirations for technological empowerment in developing countries, and particularly LDCs, in this highly complex terrain. International negotiations and developments in the context of climate change and the green economy as part of the Rio+20 framework raise several important issues for developing countries. Joining forces with the work being done by other United Nations agencies in this regard, this Report has called for the international agenda to place a greater focus on the elimination of energy poverty by promoting the greater use of RETs for mitigating climate change. This is not simply a matter of rhetoric; it implies a greater emphasis on RETs in the climate change financing architecture and the technology transfer discourse. In the context of the green economy, this Report has focused on RETs mainly, emphasizing that there is a greater need to make technologies and investment available for the development, adaptation and deployment of RETs in developing countries and LDCs, rather than imposing "green" deadlines on those countries.

(iv) The diffusion of RETs in developing countries involves much more than transferring technology hardware from one location to another. This Report, noting the complexity of technological change in different contexts, calls for targeted international support to foster RETs-related learning. Such support could include the following elements:
- an international innovation network for LDCs, with a RET focus;
- global and regional research funds for RETs deployment and demonstration;
- an international RETs technology transfer fund; and,
- an international RETs training platform.

(v) National governments in developing countries can play a pivotal role in combining conventional sources of energy with RETs in ways that will not only help reduce energy poverty, but also simultaneously promote climate-friendly solutions to development. This Report proposes that developing countries adopt a national integrated innovation policy framework to create policy incentives in national innovation policies and national energy policies for the greater use, diffusion, production and innovation of RETs.

(vi) Not all of the policy options proposed in the Report are available or applicable to all developing countries and LDCs. For the poorer countries, the ability to undertake large-scale R&D or establish significant manufacturing capacity will be constrained by the relatively small size of their domestic markets, lack of access to finance and weak institutional capacity. It may be unrealistic to expect smaller countries to become price competitive in the large-scale manufacture and distribution of RETs at least in the short and mid-term. Regional cooperation arrangements will be important for innovation and energy generation and distribution for many such smaller developing countries.

(vii) While remaining ambitious, developing countries may wish to consider focusing on adapting existing RETs to their domestic contexts and markets, potentially for off-grid applications, lowering their costs and improving their operational performance. This